T0392513

Human Performance and Situation Awareness Measures

Human Performance and Situation Awareness Measures

Authored by
Valerie Jane Gawron

CRC Press is an imprint of the
Taylor & Francis Group, an **informa** business

CRC Press
Taylor & Francis Group
6000 Broken Sound Parkway NW, Suite 300
Boca Raton, FL 33487-2742

Printed on acid-free paper

International Standard Book Number-13 978-0-367-00231-2 (Hardback)

Library of Congress Cataloging-in-Publication Data

Names: Gawron, Valerie J., author.
Title: Human performance and situation awareness measures / Valerie Jane Gawron.
Other titles: Human performance measures handbook
Description: Third edition. | Boca Raton, FL : CRC Press/Taylor & Francis Group, 2019. | Original edition published under title: Human performance measures handbook. | Includes bibliographical references and index.
Identifiers: LCCN 2018037339| ISBN 9780367002312 (hardback : acid-free paper) | ISBN 9780429019531 (ebook)
Subjects: LCSH: Human engineering--Handbooks, manuals, etc. | Human-machine systems--Handbooks, manuals, etc. | Situational awareness--Handbooks, manuals, etc.
Classification: LCC T59.7 .G38 2019 | DDC 620.8/2--dc23
LC record available at https://lccn.loc.gov/2018037339

Visit the Taylor & Francis Web site at
http://www.taylorandfrancis.com

and the CRC Press Web site at
http://www.crcpress.com

To my parents: Jane Elizabeth Gawron 12 June 1926 to 17 March 2002

and Stanley Carl Gawron 17 March 1921 to 9 February 2000.

Contents

List of Figures

List of Tables

Preface

This *Human Performance and Situation Awareness Measures* handbook was developed to help researchers and practitioners select measures to be used in the evaluation of human/machine systems. It can also be used to supplement classes at both the undergraduate and graduate courses in ergonomics, experimental psychology, human factors, human performance, measurement, and system test and evaluation. Volume 1 of this handbook begins with an overview of the steps involved in developing a test to measure human performance, workload, and/or situational awareness. This is followed by a definition of human performance and a review of human performance measures. Situational Awareness is similarly treated in subsequent chapters. Finally, workload is defined and measures described in Volume 2.

Acknowledgments

This book began while I was supporting numerous test and evaluation projects of military and commercial transportation systems. Working with engineers, operators, managers, programmers, and scientists showed a need for both educating them on human performance measurement and providing guidance for selecting the best measures for the test. I thank Dr. Dave Meister, who provided great encouragement to me to write this book based on his reading of my "measure of the month" article in the Test and Evaluation Technical Group newsletter. He, Dr. Tom Enderwick, and Dr. Dick Pew also provided a thorough review of the first draft of the first edition of this book. For these reviews I am truly grateful. I miss you, Dave.

Author

Valerie Gawron has a BA in Psychology from the State University College at Buffalo, a MA also in Psychology from the State University College at Geneseo, a PhD in Engineering Psychology from the University of Illinois, and a MS in Industrial Engineering and MBA, both from the State University of New York at Buffalo. She completed postdoctoral work in environmental effects on performance at the New Mexico State University in Las Cruces and began work for Calspan directly following. She remained at Calspan for 26 years until it was eventually acquired by General Dynamics and she was made a technology fellow. She is presently a human factors engineer at the MITRE Corporation. Dr. Gawron has provided technical leadership in Research, Development, Test, and Evaluation of small prototype systems through large mass-produced systems, managed million dollar system development programs, led the design of information systems to support war fighters and intelligence personnel, fielded computer-aided engineering tools to government agencies and industry, tested state-of-the-art displays including Helmet Mounted Displays, Night Vision Goggles, and Synthetic Vision Displays in military and commercial aircraft, evaluated security systems for airports and United States Embassies, conducted research in both system and human performance optimization, applied the full range of evaluation tools from digital models through human-in-the-loop simulation to field operational tests for military, intelligence, and commercial systems, directed accident reenactments, consulted on driver distraction, accident investigation, and drug effects on operator performance, and written over 425 publications including the *Human Performance, Workload, and Situation Awareness Measures Handbook* (second edition) and *2001 Hearts: The Jane Gawron Story*. Both are being used internationally in graduate classes, the former in human factors and the latter in patient safety.

Dr. Gawron has served on the Air Force Scientific Advisory Board, the Army Science Board, the Naval Research Advisory Committee, and the National Research Council. She gives workshops on a wide range of topics to very diverse audiences, from parachute testing given as part of the Sally Ride Science Festival for girls ages 8 to 14 to training applications of simulation to managers and engineers. She has worked programs for the United States Air Force, Army, Navy, Marines, NASA, the Departments of State and Justice, the Federal Aviation Administration, the Transportation Security Administration, the National Transportation Safety Board, the National Traffic Safety Administration, as well as for commercial customers. Some of this work has been international and Dr. Gawron has been to 195 countries. Dr. Gawron is an associate fellow of the American Institute of Aeronautics and Astronautics, a fellow of the Human Factors and Ergonomics Society, and a fellow of the International Ergonomics Association.

1

Introduction

Human factors specialists, including ergonomists, industrial engineers, engineering psychologists, human factors engineers, and many others, continually seek better (more efficient and effective) ways to characterize and measure the human element as part of the system so we can build trains, planes, and automobiles; process control stations, and other systems with superior human/system interfaces. Yet the human factors specialist is often frustrated by the lack of readily accessible information on human performance, workload, and Situational Awareness (SA) measures. To fill that void, this book was written to guide the reader through the critical process of selecting the appropriate measures of human performance, workload, and SA for objective evaluations.

There are two types of evaluations of human performance. The first type is subjective measures. These are characterized by humans providing opinions through interviews and questionnaires or by observing others' behavior. There are several excellent references on these techniques (e.g., Meister, 1986). The second type of evaluation of human performance is the experimental method. Again, there are several excellent references (e.g., Keppel, 1991; Kirk, 1995). This experimental method is the focus of this book.

Chapter 1 is a short tutorial on the experimental design. For the tutorial, the task of selecting between aircraft cockpit displays is used as an example. For readers familiar with the general principles of experimentation, this should be simply an interesting application of academic theory. For readers who may not be so familiar, it should provide a good foundation of why it is so important to select the right measures when preparing to conduct an experiment.

Chapter 2 describes measures of human performance and Chapter 3 describes measures of SA. Each measure is described, along with its strengths and limitations, data requirements, threshold values, and sources of further information. To make this desk reference easier to use, extensive author and subjective indices are provided.

1.1 The Example

An experiment is a comparison of two or more ways of doing things. The "things" being done are called *independent variables*. The "ways" of doing things are called *experimental conditions*. The measures used for comparison are *dependent variables*. Designing an experiment requires: defining the independent variables, developing the experimental conditions, and selecting the dependent variables. Ways of meeting these requirements are described in the following steps.

1.1.1 Step 1: Define the Question

Clearly define the question to be answered by the results of the experiment. Let's work through an example. Suppose a moving map display is being designed and the lead engineer wants to know if the map should be designed as track up, north up, or something else. He comes to you for an answer. You have an opinion but no hard evidence. You decide to run an experiment. Start by working with the lead engineer to define the question. First, what are the ways of displaying navigation information, that is, what are the experimental conditions to be compared? The lead engineer responds, "Track up, north up, and maybe something else." If he cannot define something else, you cannot test it. So now you have two experimental conditions: track up versus north up. These conditions form the two levels of your first independent variable, direction of map movement.

1.1.2 Step 2: Check for Qualifiers

Qualifiers are independent variables that qualify or restrict the generalizability of your results. In our example, an important qualifier is the type of user of the moving map display. Will the user be a pilot (who is used to track up) or a navigator (who has been trained with north-up displays)? If you run the experiment with pilots, the most you can say from your results is that one type of display is best *for pilots*. There is your qualifier. If your lead engineer is designing moving map displays for both pilots and navigators, you have only given him half an answer or worse, if you did not think about the qualifier of type of user, you may have given him an incorrect answer. So, check for qualifiers and use the ones that will have an effect on decision making as independent variables.

In our example, the type of user will have an effect on decision making, so it should be the second independent variable in the experiment. Also in our example, the size of the display will not have an effect on decision making since the lead engineer only has room for an 8-inch display in the instrument panel. Therefore, size of the display should not be included as an independent variable.

1.1.3 Step 3: Specify Conditions

Specify the exact conditions to be compared. In our example, the lead engineer is interested in track up versus north up. So, the movement of the map will vary between the two conditions but everything else about the displays (e.g., scale factor, display resolution, color quality, size of the display, and so forth) should be exactly the same. This way, if the participants' performance using the two types of displays is different, that difference can be attributed only to the type of display and not to some other difference between the displays.

1.1.4 Step 4: Match Participants

Match the participants to the end users. If you want to generalize the results of your experiment to what will happen in the real world, try to match the participants to the users of the system in the real world. This is extremely important since participants' past experiences may greatly affect their performance in an experiment. In our example, we added a second independent variable to our experiment specifically because of participants' previous experiences (that is, pilots are used to track up, navigators are trained with north up). If the end users of the display are pilots, we should use pilots as our participants. If the end users are navigators, we should use navigators as our participants. Other participant variables may also be important; in our example, age and training are both very important. Therefore, you should identify what training the user of the map display must have and provide that same training to the participants before the start of data collection.

Age is important because pilots in their forties may have problems focusing on near objects such as map displays. Previous training is also important: F-16 pilots have already used moving map displays while C-130 pilots have not. If the end users are pilots in their twenties with F-16 experience and your participants are pilots in their forties with C-130 experience, you may be giving the lead engineer the wrong answer to his question of which type of display is better.

1.1.5 Step 5: Select Performance Measures

Your results are influenced to a large degree by the performance measures you select. Performance measures should be relevant, reliable, valid, quantitative, and comprehensive. Let's use these criteria to select performance measures for our example problem.

Criteria 1: Relevant. Relevance to the question being asked is the prime criteria to be used when selecting performance measures. In our example, the lead engineer's question is "What type of display format is better?" Better can refer to staying on course better (accuracy) but it can also refer to getting to the waypoints on time better (time). Participants' ratings of which display format they prefer does not answer the question of which display is better from a performance standpoint because preference ratings can be affected by factors other than performance.

Criteria 2: Reliable. Reliability refers to the repeatability of the measurements. For recording equipment, reliability is dependent on careful calibration of equipment to ensure that measurements are repeatable and accurate (i.e., an actual course deviation of 50.31 feet should always be recorded as 50.31 feet). For rating scales, reliability is dependent on the clarity of the wording. Rating scales with ambiguous wording will not give reliable measures of performance. For example, if the question on the rating scale is "Was your performance okay?" the participant may respond "No" after his first simulated flight but "Yes" after his second simply because he is more comfortable with the task. If you now let him repeat his first flight, he may respond, "Yes." In this case, you are getting a different answer to the same question in the same condition. Participants will give more reliable responses to less ambiguous questions such as "Did you deviate more than 100 feet from course in this trial?" Even so, you may still get a first "No" and a second "Yes" to the more precise question, indicating that some learning had improved his performance the second time.

Participants also need to be calibrated. For example, if you are asking which of eight flight control systems is best and your metric is an absolute rating (e.g., Cooper-Harper Handling Qualities Rating), your participant needs to be calibrated with both a "good" aircraft and a "bad" aircraft at the beginning of the experiment. He may also need to be recalibrated during the course of the experiment. The symptoms that suggest the need to recalibrate your participant are the same as those that indicate that you should recalibrate your measuring equipment: (a) all the ratings are falling in a narrower band than you expect, (b) all the ratings are higher or lower than you expect, and (c) the ratings are generally increasing (or decreasing) across the experiment independent of experimental condition. In these cases, give the participant a flight control system that he has already rated. If this second rating is substantially different from the one he previously gave you for the same flight control system, you need to recalibrate your participants with an aircraft that pulls their ratings away from the average: bad aircraft if all the ratings are near the top, good aircraft if all the ratings are near the bottom.

Criteria 3: Valid. Validity refers to measuring what you really think you are measuring. Validity is closely tied to reliability. If a measure is not reliable, it can never be valid. The converse is not necessarily true. For example, if you ask a participant to rate his workload from 1 to 10 but do not define for him what you mean by workload, he may rate the perceived difficulty of the task rather than the amount of effort he expended in performing the task.

Criteria 4: Quantitative. Quantitative measures are easier to analyze than qualitative measures. They also provide an estimate of the size of the difference between experimental conditions. This is often very useful in performing trade-off analyses of performance versus cost of system designs. This criterion does not preclude the use of qualitative measures, however, because qualitative measures often improve the understanding of experiment results. For qualitative measures, an additional issue must be considered – the type of rating scale. Nominal scales assign an adjective to the system being evaluated, (e.g., easy to use). "A nominal scale is categorical in nature, simply identifying differences among things on some characteristic. There is no notion of order, magnitude or size" (Morrow et al., 1995, p. 28). Ordinal scales rank systems being evaluated on a single or a set of dimensions (e.g., the north-up is easier than the track-up display). "Things are ranked in order, but the difference between ranked positions are not comparable" (Morrow et al., 1995, p. 28).

Interval scales have equal distances between the values being used to rate the system under evaluation. For example, a bipolar rating scale is used in which the two poles are *extremely easy to use* and *extremely difficult to use.* In between these extremes are the words *moderately easy, equally easy,* and *moderately difficult.* The judgment is that there is an equal distance between any two points on the scale. The perceived difficulty difference between *extremely* and *moderately* is the same as between *moderately* and *no difference.* However, "the zero point is arbitrarily chosen" (Morrow et al., 1995, p. 28). The final type of scale is a ratio scale which possesses a true zero (Morrow et al., 1995, p. 29). More detailed descriptions of scales are presented in Baird and Noma (1978), Torgerson (1958), and Young (1984).

Criteria 5: Comprehensive. Comprehensive means the ability to measure all aspects of performance. Recording multiple measures of performance during an experiment is cheaper than setting up a second experiment to measure something that you missed in the first experiment. So, measure all aspects of performance that may be influenced by the independent variables. In our example, participants can trade off accuracy for time (e.g., cut a leg to reach a waypoint on time) and vice versa (e.g., go slower to stay on course better), so we should record both accuracy and time measures. For an example of using these and several additional criteria in air combat maneuvering, see Lane (1986).

1.1.6 Step 6: Use Enough Participants

Use enough participants to statistically determine if there is a difference in the values of the dependent variables between the experimental conditions. In our example, is the performance of participants using the track-up display versus the north-up display statistically different? Calculating the number of participants you need is very simple. First, predict how well participants will perform in each condition. You can do this using your own judgment, previous data from similar experiments, or from pretest data using your experimental setup. In our example, how much error will there be in waypoint arrival times using the track-up display and the north-up display? From previous studies, you may think that the average error for pilots using the track-up display will be 1.5 seconds and using the north-up display, 2 seconds. Similarly, the navigators will have about 2 seconds error using the track-up display and 1.5 seconds error with the north-up display. For both sets of participants and both types of displays, you think the standard deviation will be about 0.5 second.

Now we can calculate the effect size, that is, the difference between performances in each condition:

$$\text{Effect size} = \frac{|\text{performance in track up} - \text{performance in north up}|}{\text{Standard deviation}}$$

$$\text{Effect size for pilots} = \frac{|1.5 - 2|}{0.5} = 1$$

$$\text{Effect size for navigators} = \frac{|2 - 1.5|}{0.5} = 1$$

In Figure 1.1 we can now read the number of participants needed to discriminate the two conditions. For an effect size of 1, the number of participants needed is 18. Therefore, we need 18 pilots and 18 navigators in our experiment. Note that although the function presented in Figure 1.1 is not etched in stone, it is based on over 100 years of experimentation and statistics.

Note that you should estimate your effect size in the same units as you will use in the experiment. Also note that because effect size is calculated as a ratio, you will get the same effect size (and hence the same number of participants) for equivalent measures. Finally, if you have no idea of the effect size, try the experiment yourself and use your own data to estimate the effect size.

1.1.7 Step 7: Select Data-Collection Equipment

Now that you know the size of the effect of the difference between conditions, check that the data-collection equipment you have selected can reliably measure performance at least one order of magnitude smaller than the smallest discriminating decimal place in the size of the expected difference between conditions. In our example, the expected size in one condition was 1.5 seconds. The smallest discriminating decimal place (1.5 vs. 2.0) is tenths of a second. One order of magnitude smaller is hundredths. Therefore, the recording equipment should be accurate to 1/100th of a second.

1.1.8 Step 8: Match Trials

Match the experimental trials to the end usage. As in Step 4, if you want to generalize the results of your experiment to what will happen in the real world, you must match the experimental trials to the real world.

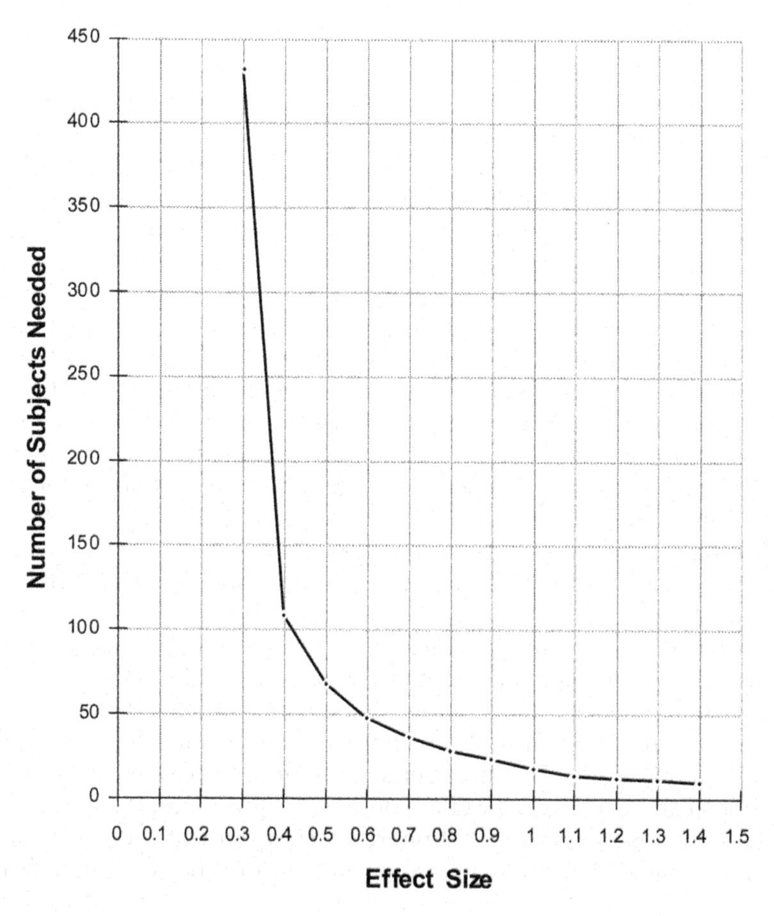

FIGURE 1.1
Number of participants needed as a function of effect size.

(Note, a single trial is defined as continuous data collection under the same experimental conditions. For example, three successive instrument approaches with the same flight-control configuration constitute one trial.) The following are important characteristics to match.

Characteristic 1: Length of the Trial. Over the length of a trial, performance improves due to warm-up effects and learning and then degrades as fatigue sets in. If you measure performance in the experiment for 10 minutes but in the real world, pilots and navigators perform the task for two hours, your results may not reflect the peak warm-up or the peak fatigue. Consequently, you may give the lead engineer the wrong answer. So always try to match the length of each experimental trial to the length of the task in the real world.

Characteristic 2: Level of Difficulty. If you make the experimental task too easy, all the participants will get the same performance score: 0 errors. If all the performance scores are the same, you will not be able to distinguish between experimental conditions. To avoid this problem, make the task realistically difficult. In general, the more difficult the task in the experiment, the more likely you are to find a statistical difference between experimental conditions. This is because difficulty enhances discriminability between experimental conditions. However, there are two exceptions that should be avoided in any experiment. First, if the experimental task is too difficult, the performance of all the participants will be exactly the same: 100% errors. You will have no way of knowing which experimental condition is better and the experiment was useless. Second, if you increase the difficulty of the task beyond that which can ever be expected in the real world, you may have biased your results. In our example, you may have found that track-up displays are better than north-up displays in mountainous terrain, flying under 100 feet Above Ground Level (AGL) at speeds exceeding 500 knots with wind gusts over 60 knots. But how are they in hilly terrain, flying at 1000 feet AGL at 200 knots with wind gusts between 10 and 20 knots, that is, in the conditions in which they will be used nearly 70% of the time? You cannot answer this question from the results of your experiment – or if you give an answer, it may be incorrect. Therefore, typical conditions should be conditions of the experiment.

Characteristic 3: Environmental Conditions. Just as in Step 4 where you tried to match the participants to the end users, you should also try to match the environmental conditions of the laboratory (even if that laboratory is an operational aircraft or an in-flight simulator) to the environmental conditions of the real world. This is extremely important because environmental conditions can have a greater effect on performance than the independent variables in your experiment. Important environmental conditions that should be matched include lighting, temperature, noise, and task load. Lighting conditions should be matched in luminance level (possible acuity differences), position of the light source (possible glare), and type of light source (incandescent lights have "hot spots" that can create point sources of glare; fluorescent lights provide even, moderate light levels; sunlight can mask some colors and create large glare spots). Temperatures above 80 degrees Fahrenheit decrease the amount of effort participants expend; temperatures below 30 degrees Fahrenheit make fine motor movements (e.g., setting radio frequencies) difficult. Noise can either enhance or degrade performance: enhancements are due to increased attention; degradations are due to distractions. Meaningful noise (e.g., a conversation) is especially distracting. Task load refers to both the number and types of tasks that are being performed at the same time as your experimental task. In general, the greater the number of tasks that are being performed simultaneously and the greater the similarity of the tasks that are being performed simultaneously, the worse the performance on the experimental task. The classic example is monitoring three radio channels simultaneously. If the volume or quality of the communications is not varied (thus making the tasks less similar), this task is extremely difficult.

1.1.9 Step 9: Select Data-Recording Equipment

In general, the data-recording equipment should be able to record data for 1.5 times the length of the experimental trial. This allows for false starts without changing the data tape, disk, or other storage medium. The equipment should be able to have separate channels for each continuous dependent variable (e.g., altitude, airspeed) and as many channels as necessary to record the discrete variables (e.g., reaction time (RT) to a simulated fire) without any possibility of recording the discrete variables simultaneously on the same channel (thus losing valuable data).

1.1.10 Step 10: Decide Participant Participation

Decide if each participant should participate in all levels of the condition in the experiment. There are many advantages of having a single participant participate in more than one experimental condition: (a) reduced recruitment costs, (b) decreased total training time, and (c) better matching of participants across experimental conditions. But there are some conditions that preclude using the same participant in more than one experimental condition. The first is previous training. In our example, pilots and navigators have had very different training. The differences in their training may affect their performance; therefore, they cannot participate in both roles: pilot and navigator. Second, some experimental conditions can make the participants' performance worse than even untrained participants in another experimental condition. This effect is called negative transfer. Negative transfer is especially strong when two experimental conditions require a participant to give a different response to the same stimulus. For example, the response to a fire alarm in Experimental Condition 1 is pull the T handles, then feather the engine. In Experimental Condition 2, the response is feather the engine and then pull the T handle. Participants who have not participated in any experimental condition are going to have faster RTs and fewer errors than participants who have already participated in either Experimental Condition 1 or 2. Whenever there is negative transfer (easy to find by comparing performance of new participants to participants who have already participated in another condition), use separate participants.

Learning is another important condition affecting the decision to use the same participants or not. Participants who participate in more than one experimental condition are constantly learning about the task that they are performing. If you plot the participants' performance (where high scores mean good performance) on the ordinate and the number of trials he/she has completed along the abscissa, you will find a resulting J curve where a lot of improvement in performance occurs in the first few trials and very little improvement occurs in the final trials. The point at which there is very little improvement is called *asymptotic learning*. Unless participants are all trained to asymptote before the first trial, their performance will improve over the entire experiment regardless of the differences in the experimental conditions. Therefore, the "improvement" you see in later experimental conditions may have nothing to do with what the experimental condition is but rather with how long the participant has been performing the task in the entire experiment.

A similar effect occurs in simple, repetitive, mental tasks and all physically demanding tasks. This effect is called *warm-up*. If the participants' performance improves over trials regardless of the experimental conditions, you may have a warm-up effect. This effect can be eliminated by having participants perform preliminary trials until their performance on the task matches their asymptotic learning.

The final condition is fatigue. If the same participant is performing more than one trial, fatigue effects may begin to mask the differences in the experimental conditions. You can check for fatigue effects in four ways: by performing a number of trials yourself (how are you feeling?); by observing your participants (are they showing signs of fatigue?); by comparing performance in the same trial number but different conditions across participants (is everyone doing poorly after three trials?); and by asking the participants how they are feeling.

1.1.11 Step 11: Order the Trials

In Step 10, we described order or carry over effects. Even if these do not occur to a great degree or if they do not seem to occur at all, it is still important to order your data-collection trials so as to minimize order and carry over effects. Another important carry over effect is the experimenter's experience – during your first trial, experimental procedures may not yet be smoothed out. By the 10th trial, everything should be running efficiently and you may even be anticipating participants' questions before they ask them. The best way to minimize order and carry over effects is to use a Latin-square design. This design ensures that every experimental condition precedes and succeeds every other experimental condition an equal number of times.

Once the Latin square is generated, check the order for any safety constraints (e.g., landing a Level 3 aircraft in maximum turbulence or severe crosswinds). Adjust this order as necessary to maintain safety. The resulting numbers indicate the order in which you should collect your data. For example, Participant 1 gets north up then track up. Participant 2 gets the opposite. Once you have completed data collection for the pilots, you can collect data on the navigators. It does not matter what order you collect the pilots' and navigators' data because the pilots' data will never be compared to the navigators' data, that is, you are not looking for an interaction between the two independent variables. If the second independent variable in the experiment had been size (e.g., the lead engineer gives you the option for an 8- or 12-inch display), the interaction would have been of interest. For example, are 12-inch, track-up displays better than 8-inch, north-up displays? If we had been interested in this interaction, a Latin square for four conditions: Condition 1, 8-inch, north up; Condition 2, 8-inch, track up; Condition 3, 12-inch, north up; and Condition 4, 12-inch, track up would have been used.

1.1.12 Step 12: Check for Range Effects

Range effects occur when your results differ based on the range of experimental conditions that you use. For example, in Experiment 1 you compare track-up and north-up displays, and find that for pilots track-up displays are better. In Experiment 2, you compare track-up, north-up, and horizontal situation indicator (HSI) displays. This time you find no difference between track-up and north-up displays but both are better than a conventional HSI. This is an example of a range effect: when you compare across one range of conditions, you get one answer; when you compare across a second range of conditions, you get another answer. Range effects are especially prevalent when you vary environmental conditions such as noise level and temperature. Range effects cannot be eliminated. This makes selecting a range of conditions for your experiment especially important.

To select a range of conditions, first return to your original question. If the lead engineer is asking which of two displays to use, Experiment 1 is the right experiment. If he is asking whether track-up or north-up displays are better than an HSI, Experiment 2 is correct. Second, you have to consider how many experimental conditions your participants are experiencing. If it is more than seven, your participant is going to have a hard time remembering what each condition was but his or her performance will still show the effect. To check for a "number of trials" effect, plot the average performance in each trial versus the number of the trials the participant has completed. If you find a general decrease in performance, it is time to either reduce the number of experimental conditions that the participant experiences or provide long rest periods.

1.2 Summary

The quality and validity of the data are improved by incorporating the following steps in the experimental design:

Step 1: Clearly define the question to be answered.

Step 2: Check for qualifiers.

Step 3: Specify the exact conditions to be compared.

Step 4: Match the participants to the end users.

Step 5: Select performance measures.
Step 6: Use enough participants.
Step 7: Select data-collection equipment.
Step 8: Match the experimental trials to the end usage.
Step 9: Select data-recording equipment.
Step 10: Decide if each participant should participate in all levels.
Step 11: Order the trials.
Step 12: Check for range effects.

Step 5 is the focus for the remainder of this book.

Sources

Baird, J.C., and Noma, E. *Fundamentals of Scaling and Psychophysics*. New York: Wiley, 1978.

Keppel, G. *Design and Analysis: A Researcher's Handbook*. Englewood Cliffs, NJ: Prentice Hall, 1991.

Kirk, R.R. *Experimental Design: Procedures for the Behavioral Sciences*. Pacific Grove, California: Brooks/Cole Publishing Company, 1995.

Lane, N. *Issues in Performance Measurement for Military Aviation with Applications to Air Combat Maneuvering (NTSC TR-86-008)*. Orlando: Naval Training Systems Center, April 1986.

Meister, D. *Human Factors Testing and Evaluation*. New York: Elsevier, 1986.

Morrow, J.R., Jackson, A.W., Disch, J.G., and Mood, D.P. *Measurement and Evaluation in Human Performance*. Champaign, IL: Human Kinematics, 1995.

Torgerson, W.S. *Theory and Methods of Scaling*. New York: Wiley, 1958.

Young, F.W. Scaling. *Annual Review of Psychology* 35: 55–81, 1984.

2

Human Performance

Human performance is the accomplishment of a task by a human operator or by a team of human operators. Tasks can vary from the simple (card sorting) to the complex (landing an aircraft). Humans can perform the task manually or monitor an automated system. In every case, human performance can be measured and this handbook can help.

Performance measures can be placed into six categories. The first is accuracy in which the measure is assessing the degree of correctness. Such measures assume that there is a correct answer. Human performance measures in this category are presented in Section 2.1. The second category of human performance measures is time. Measures in this category assume that tasks have a well-defined beginning and end so that the duration of task performance can be measured. Measures in this category are listed in Section 2.2. The third category is task battery. Task batteries are collections of two or more tasks performed in series or in parallel to measure a range of abilities or effects. These batteries assume that human abilities vary across types of tasks or are differentially affected by independent variables. Examples are given in Section 2.3.

The fourth category of human performance measures is domain-specific measures which assess abilities to perform a family of related tasks. These measures assume that performance vary across segments of a mission or on the use of different controllers. Examples in this category are presented in Section 2.4. The fifth category is critical incidents which is typically used to assess worst case performance (see Section 2.5). The final category is team performance measures. These assess the abilities of two or more persons working in unison to accomplish a task or tasks. These measures assume that human performance varies when part of a team. Team performance measures are described in Section 2.6. There are also measures that fall into more than one category. This is true for all of the team measures (Section 2.6).

For uniformity and ease of use, each discussion of a measure of human performance has the same sections:

1. General description of the measure;
2. Strengths and limitations or restrictions of the measure, including any known proprietary rights or restrictions as well as validity and reliability data;
3. Data collection, reduction, and analysis requirements;

4. Thresholds, the critical levels of performance above or below that the researcher should pay attention; and
5. Sources of further information and references.

2.1 Accuracy

The first category of human performance measures is accuracy in which the measure is assessing the degree of correctness. Such measures assume that there is a correct answer.

General description – Accuracy is a measure of the quality of a behavior. Measures of accuracy include correctness score (Section 2.1.3), number correct (Section 2.1.7), percent correct (Section 2.1.9), percent of correct detections (Section 2.1.10), and probability of correct detections Section (2.1.12).

Error can also be used to measure accuracy – or the lack thereof. Error measures include absolute error (Section 2.1.1), average range scores (Section 2.1.2), deviations (Section 2.1.4), error rate (Section 2.1.5), false alarm rate (Section 2.1.6), number of errors (Section 2.1.8), percent errors (Section 2.1.11), and root mean square error (Section 2.1.14). Errors can be of omission (i.e., leaving a task out) or commission (i.e., doing the task but not correctly). Further, Reichenbach et al. (2010) reported three types of commission error: (1) incomplete cross checks, (2) discounting of contradictory information, and (3) looking but not seeing. Helton and Head (2012) reported significant increases in errors of omission after a 7.1 magnitude earthquake but errors of commission were dependent on individual differences in stress response. Finally, sometimes accuracy and error measures are combined to provide ratios (Section 2.1.13).

Strengths and limitations – Accuracy can be measured on a ratio scale and is, thus, mathematically robust. However, distributions of the number of errors or the number correct may be skewed and, thus, may require mathematical transformation into a normal distribution. In addition, some errors rarely occur and are, therefore, difficult to investigate (Meister, 1986). There is also a speed-accuracy tradeoff that must be considered (Drinkwater, 1968). Data collection requirements as well as thresholds are discussed for each accuracy measure. Source information is presented at the end of Section 2.1.

2.1.1 Absolute Error

Mertens and Collins (1986) used absolute and root mean square error on a two-dimensional compensatory tracking task to evaluate the effects of age (30 to 39 versus 60 to 69 years old), sleep (permitted versus deprived), and altitude (ground versus 3,810 m). Performance was not significantly affected by age but was significantly degraded by sleep deprivation and altitude. Similar results occurred for a problem-solving task.

Elvers et al. (1993) reported that as the probability of a task requiring the participant to determine a volume rather than a distance increased, absolute error of the distance judgment increased.

2.1.2 Average Range Score

Rosenberg and Martin (1988) used average range scores ("largest coordinate value minus smallest coordinate value," p. 233) to evaluate a digitizer puck (i.e., a cursor-positioning device for digitizing images). There was no effect of type of optical sight, however, magnification improved performance.

2.1.3 Correctness Score

Correctness scores were developed to evaluate human problem solving. A score using the following five-point rating scale was awarded based on a participant's action:

0 – Participant made an incorrect or illogical search.
1 – Participant asked for information with no apparent connection to the correct response.
2 – Participant asked for incorrect information based on a logical search pattern.
4 – Participant was on the right track.
5 – Participant asked for the key element (Griffin and Rockwell, 1984).

This measure requires well-defined search patterns. The thresholds are 0 (poor performance) to 5 (excellent performance).

Griffin and Rockwell (1984) used correctness scores to measure a participant's problem-solving performance. These authors applied stepwise regression to predict correctness scores for four scenarios: "(1) an oil-pressure gauge line break, (2) a vacuum-pump failure, (3) a broken magneto drive gear, and (4) a blocked static port" (p. 575). Demographic data, experience, knowledge scores, and information-seeking behavior were only moderately related to correctness scores. The participants were 42 pilots with 50 to 15,000 flight hours of experience.

2.1.4 Deviations

Deviations are usually measured in time or distance. For time, Ash and Holding (1990) used timing accuracy (i.e., difference between the mean spacing between notes played and the metronome interval) to evaluate keyboard-training methods. There was a significant training method effect but no significant trial or order effects on this measure.

For distance, Yeh and Silverstein (1992) asked participants to make spatial judgments of simplified aircraft landing scenes. They reported that participants were less accurate in making altitude judgments relative to depth judgments. However, altitude judgments (in mean percent correct) were more accurate as altitude increased and with the addition of binocular disparity.

In another aviation distance deviation study, McClernon et al. (2012) introduced a variable error measure to evaluate performance of 20 students with no flying experience performing a flight task using a desk-top simulator. The task was transitioning between issued clearances while maintaining 10-degree pitch and 20-degree roll. Half of the participants received stress training including placing a foot in a cold pressor. Both root mean square error (rmse) and variance were calculated for altitude, heading, pitch, and toll. The stress group had significantly lower rmse altitude and variances for both heading and pitch.

In a ground distance deviation study, Van der Kleij and te Brake (2010) studied the effects of map characteristics on the coordinated navigation of two persons to a specified point. These authors used distance between the location indicated by a participant as the navigation point and the actual location of the navigation point on a digitized map. They reported decreased deviations when the maps of the two team members were in the same orientation than when the maps were not. Further, map grids also decreased the deviations to the navigation point. There were no differences, however, between use of landmark and compass rose enhanced maps in the non-identical map condition.

2.1.5 Error Rate

Error rate has been defined as the number of incorrectly answered and unanswered problems divided by the total number of problems presented. Wierwille et al. (1985) reported that error rate was significantly related to the difficulty of a mathematical problem-solving task. Error rates were higher for rapid communication than for conventional visual displays (Payne and Lang, 1991). In a similar study, Cook et al. (2010) reported significant differences in error rates between 2D visual displays and augmented 2D or 3D visual display types for replanning Unmanned Aerial Vehicle (UAV) routes.

Eatchel et al. (2012) reported a significant increase in error rate in a memory task when unrelated images were used to interrupt the task than when the task was performed without interruptions. Li and Oliver (2000) reported significant differences in error rate for drivers identifying roads from electronic maps at different levels of road complexity, map orientation, and the interaction of these two independent variables.

Error rates did not discriminate between observers exposed to strobe lights and those who were not exposed (Zeiner and Brecher, 1975). Nor were error rates significantly different among five feedback conditions (normal, auditory, color, tactile, and combined) for computer mouse use (Akamatsu et al., 1995). Akamatsu et al. (1995) reported no significant differences in error rate among tactile, auditory, and visual feedback pointing systems.

In another measure of error rate, Kopardekar and Mital (1994) reported normalized errors (number of errors divided by time) were significantly different for work break schedules. The fewest errors for directory assistance operators occurred in the 120 minutes of continuous work with no breaks than the 30 minutes with five-minute breaks or the 60-minute work with 10-minute breaks. In still another related measure, St. John and Risser (2009) reported significant differences in the rate of missed detections of changes in a visual presentation of a truck.

2.1.6 False Alarm Rate

False alarm rate is typically used in sensory perception studies.

Auditory perception – Mullin and Corcoran (1977) reported that the false alarm rate decreased over time in an auditory vigilance task. There was no significant effect of amplitude of the auditory signal or time of day (08:30 versus 20:30).

Visual perception – Studies in visual perception evaluate characteristics of the visual stimulus including number, complexity, as well as observer characteristics (e.g., time on task, emotion).

Lanzetta et al. (1987) reported that the false alarm rate significantly decreased as the presentation rate increased (9.5%, 6/minute; 8.2%, 12/minute; 3.2%, 24/minute; 1.5%, 48/minute). In a similar study, Teo and Szalma (2010) reported a significantly higher false alarm rate in an 8 visual displays condition than in the 1, 2, or 4 visual displays conditions.

Loeb et al. (1987) found the number of false alarms was significantly lower in a simple than a complex task. Swanson et al. (2012) reported lower false alarm rates for low rather than high Ground Sample Distance, long than short dwell time, and long rather than short aspect angle. The task was performing Intelligence, Surveillance, and Reconnaissance from Remotely Piloted Aircraft imagery.

Galinsky et al. (1990) used false alarm rate to evaluate periods of watch (i.e., five 10-minute intervals). False alarm rate significantly increased as event rate decreased (five versus 40 events per minute).

Colquhoun (1961) reported that 15 of 21 participants in a visual detection task had no false alarms while the remaining six varied considerably in the false alarm rate. There was a trend for false alarm rate to decrease over time. There was a significant positive correlation between number of false alarms and number of detected signals. Funke et al. (2016) reported that observers who were paired but acting independently detected significantly more signals in a visual vigilance task than observers working alone.

In an unusual study, Culley et al. (2011) reported that false alarm rate was significantly affected by anger, fear, and sorrow. Their participants were 100 undergraduate students performing a simulated airport luggage inspection task. Those in the anger group (manipulated by a CNN report after which participants were asked what made them so angry) had a significantly higher false alarm rate than those in the fear group (same CNN report but subsequent question asked what made you so afraid) or the sorrow group (same manipulation but question was about sorrow).

2.1.7 Number Correct

Similar to false alarm rate, number correct is also affected by stimulus characteristics and observer characteristics.

Stimulus Characteristics. Tzelgov et al. (1990) used the number correct to evaluate two types of stereo picture compression. This measure was significantly different between tasks (higher in object decision task than in-depth decision task), depth differences (greater at larger depth differences), and presentation condition. There were also significant interactions. Curtis et al. (2010) reported no significant difference in the number correct in a task to determine if two aircraft were the same when viewed at different slant angles with no feedback, knowledge of results, or conceptual feedback.

Loeb et al. (1987) found no significant effects of cueing, constant or changing target, or brief or persistent target on number correct. In contrast, McDougald and Wogalter (2011) reported increased number correct when relevant parts of warning pictorials were highlighted.

Observer Characteristics. Smith and Miles (1986) reported a decrease in the number of targets detected for participants who ate lunch prior to the task as compared to participants who did not eat lunch prior. Fleury and Bard (1987) reported that the number correct in a visual detection task was decreased after aerobic effort as compared to pretest inactivity. Raaijmaker and Verduyn (1996) used the number of problems correctly solved on a fault diagnosis task to evaluate a decision aid. In an unusual application, Millar and Watkinson (1983) reported the number of correctly recognized words from those presented during surgery beginning after the first incision.

Van Orden et al. (1996) used average number of correct to evaluate the effect of cold stress on a command and control task. There were no significant differences.

In a combined stimulus and observer study, Craig et al. (1987) reported a significant decrease in the number of correct detections as event rate increased, signal frequency decreased, time on task increased, and the stimulus degraded.

For tasks with low error rates, the number completed has been used instead of the number correct. One example is the Serial Search Task in which the participant looks for pairs of digits occurring among a page full of digits (Harris and Johnson, 1978). Another example is the proportion of times that operators are able to drive a robot through an aperture. Armstrong et al. (2015) reported that as aperture size increased, the proportion of passes increases. The proportion also increased with practice.

2.1.8 Number of Errors

Number of errors has been sensitive to stimuli, participant attributes, task, and the environment.

2.1.8.1 Effects of Stimuli Characteristics on Number of Errors

In an early study, Baker et al. (1960) reported increases in the number of errors in a visual target detection task as the number of irrelevant items increased and the difference between the resolution of the reference and the target increased. Similarly, Moseley and Griffin (1987) reported a larger number of reading errors for characters with high spatial complexity.

Downing and Saunders (1987) reported significantly more errors were made in a simulated control room emergency with a mirror than with a non-mirror image control panel layout. In an aviation maintenance experiment, Warren et al. (2013) reported that Aviation Maintenance Technicians failed to investigate potential errors in a previous shift if the information was provided in written rather than face-to-face form. Frankish and Noyes (1990) used the number of data entry errors to evaluate four types of feedback: (1) concurrent visual feedback, (2) concurrent spoken feedback, (3) terminal visual feedback, and (4) terminal spoken feedback. There were no significant differences.

There have also been significant differences in number of errors related to modality. Ruffell-Smith (1979) asked participants to solve visually or aurally presented mathematical equations without using paper, calculator, or computer. The number of errors was then used to compare the performance of 20 three-person airline crews on a heavy versus a light two-leg flight. There were more errors in computation during the heavy workload condition.

Borghouts et al. (2015) used number of errors to compare infusion pump displays. Thomas et al. (2015) used number of errors to compare arrow configurations on a laptop keyboard.

2.1.8.2 Effects of Participant Characteristics on Number of Errors

Chapanis (1990) used the number of errors to evaluate short-term memory for numbers. He reported large individual differences (71 to 2,231 errors out of 8,000 numbers). He also reported a significant serial position effect (70% of the participants made the greatest number of errors at the seventh position). Women made significantly more errors than men did. In an age experiment,

Lee et al. (2010) reported that older participants (>55 years of age) made more errors in a map search task than younger participants (<55 years of age). In another age study, Mertens and Collins (1986) reported that performance was not related to age (30 to 39 versus 60 to 69 years old). However, performance degraded as a function of sleep deprivation and altitude (0 versus 3,810 m).

Casali et al. (1990) used the number of uncorrected errors to evaluate a speech recognition system. There were significant recognition accuracy and available vocabulary effects but not a significant age effect.

Vermeulen (1987) used the number of errors to evaluate presentation modes for a system-state identification task and a process-control task. There was no mode effect on errors in the first task; however, inexperienced personnel made significantly more errors than experienced personnel. For the second task, the functional presentation mode was associated with fewer errors than the topographical presentation. Again, inexperienced personnel made more errors than experienced personnel.

Billings et al. (1991) reported significant increases in the number of errors pilots made during simulated flight after alcohol dosing (blood alcohol levels of 0.025, 0.05, and 0.075).

2.1.8.3 Effects of Task Characteristics on Number of Errors

Casner (2009) reported that pilots made significantly fewer errors during the en route phase of flight than during set up, approach, or missed approach. When comparing Very High Frequency (VHF) omnidirectional range (VOR) and Global Positioning Systems (GPS), there was a significant increase in errors during setup and approach but fewer errors during the missed approach.

2.1.8.4 Effects of Environment Characteristics on Number of Errors

Kuller and Laike (1998) reported an increase in the number of proof reading errors for individuals with high critical flicker fusion frequency when reading under conventional fluorescent lighting ballasts rather than high frequency fluorescent ballasts. Dodd et al. (2014) reported a significant increase in the number of data entry errors when turbulence was added to a flight simulator.

Enander (1987) reported an increase in the number of errors in moderate (+5 degrees Centigrade) cold. However, Van Orden et al. (1996) used the average number of incorrect responses to evaluate the effect of cold stress on a command and control task. There were no significant differences. There have been inconsistent results with other stressors as well. McCann (1969) found no effect of noise on the number of errors but did find a significant increase in the number of omissions when intermittent noise was introduced into an audio-visual checking task.

2.1.9 Percent Correct

Percent correct has been used to evaluate the effects of environmental stressors, visual display characteristics, tactile display characteristics, decision aids, vigilance, task, and training.

2.1.9.1 Effects of Environmental Stressors on Percent Correct

Harris and Johnson (1978) asked participants to count the frequency of visual flashes from three light sources while being exposed to 65, 110, or 125 decibels (dB) sound for 15 or 30 minutes. There was no effect of sound on percent correct.

Harris and Shoenberger (1980) used percent correct on a counting task to assess the effects of noise (65 or 100 dBA) and vibration (0 or 0.36 Hz). There were significant main and interactive effects of these two environmental stressors. For the interaction, performance was worse in 65 dBA than 100 dBA in vibration. Performance was worse in 100 dBA when vibration was not present.

Lee and Fisk (1993) reported extremely small changes (1 to 290) in percent correct as a function of the consistency in targets in a visual search task. Harris and Johnson (1978) asked participants to count the frequency of visual flashes from three light sources while being exposed to 65, 110, or 125 dB sound for 15 or 30 minutes. There was no effect of sound on percent correct.

2.1.9.2 Effects of Visual Display Characteristics on Percent Correct

Imbeau et al. (1989) used the percent of correct answers to evaluate instrument panel lighting in automobiles. They reported that accuracy decreased as character size decreased.

Coury et al. (1989) reported that percent correct was significantly greater for bar graph displays than for digital or configurable displays after extended practice (eight trial blocks).

Matthews et al. (1989) reported a significantly lower percent of correct responses for red on green displays (79.1%) and red on blue displays (75.5%) than for monochromatic (green, red, or blue on black), achromatic (white on black), or blue on green displays (85.5%). In addition, performance was significantly worse on the first (79.8%) and last (81.8%) half hours than for the middle three hours (83.9%). In another time on task study, Teo and Szalma (2011) reported no significant effect of period on watch, however, there was a significant effect of sensory (signals same or different size) versus cognition (signals same or different value) condition with the proportion of correct detections being significantly higher in the former than the later.

Brand and Judd (1993) reported a significantly lower percent of correct responses for keyboard entry as the angle of the hard copy of which they were to enter by keyboard increased (89.8% for 90 degrees, 91.0% for 30 degrees, and 92.4% for 12 degrees). Experienced users had a significantly higher percent of correct responses (94%) than naive users (88%).

Tullis (1980) used percent correct to evaluate four display formats (narrative text, structured text, black and white graphics, and color graphics). There was no significant difference in percent correct. There was, however, a significant difference in mean response time. Chen and Tsoi (1988) used the percent of correct responses to comprehension questions to evaluate readability of computer displays. Performance was better in the slow (100 words per minute) than in the fast (200 words per minute) condition. It was also better when there were jumps of one rather than five or nine character spaces. But there was no significant difference between 20- or 40-character windows. However, Chen et al. (1988) reported *no* significant effect of jump length but a significant effect of window size. Specifically, there was a higher comprehension score for the 20-character than for the 40-character window. The significant interaction indicated that this advantage only occurred in the one-jump condition.

Rhodes et al. (2009) reported a significantly higher percent correct in a cardinal direction task with a track up rather than a north up map. Finomore et al. (2010) used percent correct and verbal response accuracy to evaluate signal difficulty (easy and hard) and communication format (radio, 3D audio, chat, and a multi modal communication tool). There were significant effects of both independent variables on both dependent variables.

Swanson et al. (2012) used a similar measure, correct detection rate, to evaluate imagery interpretation from Remotely Piloted Vehicles. The rate was higher for low than high Ground Surface Distance, long than short dwell time, and long than short inter-event time. There was no significant effect of aspect angle.

Kirschenbaum et al. (2014) reported a higher percent correct for nonexpert submarine officers when uncertainty in sonar returns was presented spatially than in tabular form.

2.1.9.3 Effects of Tactile Display Characteristics on Percent Correct

Wang et al. (2014) used percent correct to compare locations of vibration source placement (finger/hand, forearm, back, waist, thigh). They reported no difference in accuracy due to placement. Brill et al. (2015) reported higher accuracy for tactile and adiotactile cues than for 3D audio cues in locating a helicopter.

2.1.9.4 Effects of Decision Aids on Percent Correct

Adelman et al. (1993) used the mean percent of correct responses to evaluate operator performance with varying types of expert systems. The task was simulated in-flight communication. The expert systems were of three types: (1) with rule-generation capability, (2) without rule-generation capability, and (3) totally automated. The operator could screen, override, or provide a manual response. Rule-generation resulted in significantly better performance than the no rule generating capability.

2.1.9.5 Effects of Vigilance on Percent Correct

Lanzetta et al. (1987) reported a significantly higher percentage of correct detections in a simultaneous (78%) than in a successive (69%) vigilance task. In addition, the percentage correct generally decreased as the presentation rate increased (79%, 6/minute; 82%, 12/minute; 73%, 23/minute; 61%, 48/minute). This difference was significant. Percentage correct also significantly decreased as a function of time on watch (87%, 10 minutes; 75%, 20 minutes; 68%, 30 minutes; 65%, 40 minutes).

Galinsky et al. (1990) used percentage of correct detections to evaluate periods of watch (i.e., five 10-minute intervals). This percentage decreased as event rate increased from five to 40 events per minute in two (continuous auditory and continuous visual) conditions but not in a third (sensory alternation) condition. In an unusual application, Doll and Hanna (1989) forced participants to maintain a constant percent of correct responses during a detection task.

2.1.9.6 Effects of Task on Percent Correct

Kennedy et al. (1990) used percent correct on four diverse tasks (pattern comparison, grammatical reasoning, manikin, and short-term memory) to assess the effects of scopolamine and amphetamine. They found significant drug effects on all but the grammatical reasoning task although the effect on it approached statistical significance ($p < 0.0683$).

Kimchi et al. (1993) reported a significantly higher percent of correct responses on a locally-directed than a globally-directed task in a divided attention condition. There were no significant attention effects on percent correct on the globally-directed task.

Arnaut and Greenstein (1990) reported no significant difference in the percentage of control responses resulting in errors as a function of level of control input. However, Kidwell et al. (2012) reported a nearly significant difference in mean time to complete an imagery analysis task in adaptable automation mode (level of automation directly managed by the operator) rather than adaptive automation mode (level of automation managed by the automation based on operator performance).

2.1.9.7 Effects of Training on Percent Correct

Fisk and Hodge (1992) reported a significant decrease in percent correct in a visual search task after 30 days in only one of five groups (the same category and exemplars were used in training). There were no differences for new, highly related, moderately related, or unrelated exemplars.

Briggs and Goldberg (1995) used percent of correct recognition of armored tanks to evaluate training. There were significant differences in presentation time (longer times were associated with higher accuracies), view (flank views were more accurate than frontal views), model (M1 had the highest accuracy, British Challenger had the worst accuracy), and participants. There were no significant effects of component shown or friend versus foe.

Taylor and Szalma (2009) reported a significant effect of block on percent correct indicating enhanced performance over time. The task was target identification.

2.1.10 Percent Correct Detections

Pepler (1958) reported an increase in the number of missed signals as temperature increased from 67 to 92 degrees Fahrenheit. In an early study, O'Hanlon et al. (1965) reported no decrement in detecting auditory tones over a 90-minute trial. In a visual time on task experiment, Schmidt et al. (2012) reported significant increases in the proportion of correct detections over time in a vigilance task. Also, Meuter and Lacherez (2016) collected percent correct detections of fictitious threat items from 170 security screeners at an Australian international airport. There was a significant interaction in which accuracy was highest early in the shift than late in the shift but only when event rates were high but not when they were low.

A number of studies have examined the effects of visual target characteristics. Chan and Courtney (1998) reported a significant improvement in the percent of peripheral visual targets presented when the targets were magnified as compared with when they were not magnified. The difference was especially apparent at high levels of eccentricity (i.e., 3.5 and 5.0) than at low levels of eccentricity (1.0, 1.6). Donderi (1994) used percentage of targets detected to evaluate types of search for life rafts at sea. The daytime percent of correct detections were positively correlated with low contrast visual acuity and negatively correlated with vision test scores. Chong and Triggs (1989) used the percent of correct responses to evaluate the effects of type of windscreen post on target detections. There were significantly smaller percentages of correct detections for solid or no posts than for open posts.

Christensen et al. (1977) used a visual reaction time (RT) task to evaluate the effects of hypoxia. In their study 10 participants performed the task while exposed to filtered air, carbon monoxide, low oxygen, and carbon monoxide combined with low oxygen. Participants had significantly more correct detections breathing filtered air than while breathing the combination of carbon monoxide and low oxygen. The lowest percentage of correct detections occurred for the low oxygen condition. In addition, there was a significant decrease in the percent correct detections over the first 15-minute interval of a two-hour performance period.

A few studies have also examined the effects of automation and decision aids. Manzey et al. (2009) reported significantly percent correct detections when an automated decision aid was used than when it was not. Reinerman-Jones et al. (2011) used percent correct to evaluate adaptive automation in unmanned ground vehicles. They reported significantly lower percent correct detection of scene changes in the eight events per minute condition than in the four events per minute condition. Similar results occurred with lower automation than high automation. In an experiment evaluating a more advanced automation (supervisory control of multiple robots), Chen and Barnes (2012) reported that persons with higher spatial ability detected more targets than persons with lower spatial ability.

In an unusual technology evaluation, Walker et al. (2013) reported higher percent correct during the use of spearcons (speech-based earcons) than traditional methods of navigating menus.

Several studies have used detection rate. Chan et al. (2014) reported significant differences in detection rate of errors in Chinese script for number of lines displayed (2, 4, or 8) and line spacing (single, 1.5, double spacing). The highest detection rate was for four-line passages and the lowest for eight-line passages. For spacing, detection rates for 1.5 and double spacing were significantly higher than for single spacing. In another detection rate study, Teo et al. (2014) reported that knowledge of results during training resulted in significantly higher detection rates than when no knowledge of results was given.

2.1.11 Percent Errors

Percent errors has been used extensively in visual perception experiments. For example, Hancock and Caird (1993) reported significant increases in the percent of errors as the shrink rate of a target decreased. The greatest percent of errors occurred for paths with 4 steps rather than 2, 8, or 16 steps. Salame and Baddeley (1987) reported increases in percent errors as the serial position of visually presented digit sequences increased through 7 and then declined through 9. Maddox and Turpin (1986) evaluated percent errors made using mark-sensed forms. Error categories were: (1) multiple entries in the same row or column, (2) substituting an incorrect number, (3) transposing two or more numbers, and (4) omitting one or more numbers. None of these measures were affected by number orientation (horizontal or vertical), number ordering (bottom-to-top or top-to-bottom), or handedness of users (Maddox and Turpin, 1986). The range of percent errors was 2.2% to 6.3% with an average of 4% per participant (Maddox and Turpin, 1986). Percent of total errors by error category were: (1) 69.3%, (2) 17.3%, (3) 12%, and (4) 1.4% (Maddox and Turpin, 1986).

2.1.12 Probability of Correct Detections

The probability of correct detections as well as the related hit rate measures have been used in numerous perception studies.

In an auditory perception study, the probability of correct detections of target words was significantly higher for natural speech than for synthetic speech (Ralston et al., 1991). Using the same measure, Zaitzeff (1969) reported greater cumulative target acquisition probability for two- than for one-person crews. In a visual perception study, there was a significant interaction with best performance in the one display 8 events/minute and the worst in the eight displays 20 events per minute condition based on the proportion of visual signals detected (Teo and Szalma, 2010).

In a training study, Teo et al. (2013) compared two training groups – one received knowledge of results, one that did not. The group that received the knowledge of results had a significantly higher proportion of correct detections.

Using hit rate, Culley et al. (2011) evaluated the effect of emotional priming on visual threat detection. Their participants were 100 undergraduate students performing a simulated airport luggage screening task. They reported no significant main effects but there were significant interactions of trial block and affective state. Hit rate was highest in block two (of two) for anger and next highest for fear. Also using hit rate, Vuckovic et al. (2013) reported significant differences in hit rate among four air traffic control tasks. In static target-pair judgments of potential conflicts, radar was significantly better than radar with a multi-conflict display. There was no significant difference between them for a static one-target search or a static multi-pair search. For the dynamic multi-pair search task, however, the radar plus multi-conflict display had a significantly higher hit rate than radar alone.

There was a statistically significant main effect for probability of detecting UAV and Unmanned Ground Vehicles (UGVs) in a virtual scene with the current system, visual system, or visual with tactile system with the highest probability associated with the combined visual and tactile system (Oron-Gilad et al., 2015). In a medical study, Siah et al. (2015) reported a significantly higher hit rate of detecting polyps viewing colonoscopy videos for experts than novice colonoscopists.

2.1.13 Ratio of Number Correct/Number Errors

Ash and Holding (1990) used number of errors divided by number of correct responses to evaluate keyboard-training methods. They reported significant learning between the first and third trial blocks, significant differences between training methods, and a significant order effect.

2.1.14 Root Mean Square Error

Root mean square error (rmse) has been used extensively to evaluate differences among types of tracking tasks, age, display characteristics, and environmental effects.

Tracking tasks. In an early study, Kvalseth (1978) reported no significant difference in rmse between pursuit and compensatory tracking tasks. Pitrella and Kruger (1983) developed a tracking performance task to match participants in tracking experiments in rmse. Eberts (1987) used rmse on a second-order, compensatory tracking task to evaluate the effects of cueing. Cueing significantly decreased error. Vidulich (1991) reported test-retest reliability of rmse on a continuous-tracking task of +0.945. The reliability of percent correct was +0.218. Potter and Singhose (2014) used rmse to compare control elements (i.e., integrator, integrator with damping, integrator with flexible mode) during performance of a manual pursuit tracking task and reported significant differences among the elements.

Age. Wild-Wall et al. (2011) reported significantly higher rmse for older than for younger drivers in a simulated driving task.

Display characteristics. Cohen et al. (2001) reported no significant differences in rmse on a compensatory tracking task using three different types of displays: (1) compressed pitch ladder, (2) aircraft viewed from the rear, and (3) aircraft viewed from the side. More recently, Armentrout et al. (2010) reported significant differences in rmse in N1 (engine low pressure) estimation as a function of display configuration.

Environmental effects. Rmse has also been used to assess environmental effects. For example, Frazier et al. (1982) compared rmse for roll-axis tracking performance while participants were exposed to gravitational (G) forces. Rmse increased 19% from the 1 G when participants were exposed to +5 G_z for 95 seconds, 45% when exposed to combined +5G_z/±1 G_y, and 70% for +5 G_z/± G_y. In another G study, Albery and Chelette (1998) used rmse scores on tracking a visual target while exposed to +5 to +9 Gz wearing one of six G-suits. There were no significant differences among the six suits on rmse scores. Pepler (1958) used time and distance off target to assess the effect of temperature and reported lagging, overshooting, leading, and approaching effects on tracking over temperatures of 76 degrees Fahrenheit. He tested up to 91 degrees. Bohnen and Gaillard (1994) reported significant effects of sleep loss on tracking performance. Gerard and Martin (1999) used the number of contacts of a ring suspended on a rod with a wire as a measure of tracking performance. They reported that previous vibration exposure resulted in increased number of tracking errors.

Limitations. Although rmse has been used extensively, the measure is not without its critics. For example, Hubbard (1987) argued that rmse is not an adequate measure of pilot performance since deviations in one direction (e.g., down in altitude) do not have the same consequence as in the opposite direction (e.g., up in altitude).

Other measures of tracking errors have been used such as an integral of vertical and lateral error over time (Torle, 1965). This measure was used to investigate the effects of backlash, friction, and presence of an arm rest in an aircraft simulator. It was sensitive to all three independent variables.

Beshir et al. (1981) developed a tracking error score that was significantly degraded by increases in ambient temperature. Beshir (1986) used time on target (in seconds) per minute and reported decreased performance after 15 minutes on task and at 18.3 degrees Centigrade than at 21.1 degrees Centigrade.

McLeod and Griffin (1993) partitioned the total rmse into a component linearly correlated with movement of the target (input correlated error) and a component not linearly correlated with target movement (remnant). There was a significant session (more in session 1 than in session 2) as well as a duration effect (increase over 18 minutes) but no effect due to vibration of 1 octave centered on 4 Hz.

Data requirements – All correct answers must be identified prior to the start of the experiment. Errors should be reviewed to ensure that they are indeed errors and not alternative versions of the correct answers.

Thresholds – During data reduction, negative numbers of errors or of correct answers should be tested for accuracy. Percent correct or percent errors greater than 100% should also be tested for accuracy.

Sources

Adelman, L., Cohen, M.S., Bresnick, T. A., Chinnis, J.O., and Laskey, K.B. Real-time expert system interfaces, cognitive processes, and task performance: An empirical assessment. *Human Factors* 35(2): 243–261, 1993.

Akamatsu, M., MacKenzie, I.S., and Hasbroucq, T. A comparison of tactile, auditory, and visual feedback in a pointing task using a mouse-type device. *Ergonomics* 38(4): 816–827, 1995.

Albery, W.B., and Chelette, T.L. Effect of G suit on cognitive performance. *Aviation, Space, and Environmental Medicine* 69(5): 474–479, 1998.

Armentrout, J., Hansen, D., and Hall, T. Performance based selection of engine display features. Proceedings of the Human Factors and Ergonomics Society 54th Annual Meeting, 65–69, 2010.

Armstrong, M.E., Jones, K.S., and Schmidlin, E.A. Tele-operating USAR robots: Does driving performance increase with aperture width or practice? Proceedings of the Human Factors and Ergonomics Society 59th Annual Meeting, 1372–1376, 2015.

Arnaut, L.Y., and Greenstein, T.S. Is display/control gain a useful metric for optimizing an interface? *Human Factors* 32(6): 651–663, 1990.

Ash, D.W., and Holding, D.H. Backward versus forward chaining in the acquisition of a keyboard skill. *Human Factors* 32(2): 139–146, 1990.

Baker, C.A., Morris, D.F., and Steedman, W.C. Target recognition on complex displays. *Human Factors* 2(2): 51–61, 1960.

Beshir, M.Y. Time-on-task period for unimpaired tracking performance. *Ergonomics* 29(3): 423–431, 1986.

Beshir, M.Y., El-Sabagh, A.S., and El-Nawawi, M.A. Time on task effect on tracking performance under heat stress. *Ergonomics* 24(2): 95–102, 1981.

Billings, C.E., Demosthenes, T., White, T.R., and O'Hara, D.B. Effects of alcohol on pilot performance in simulated flight. *Aviation, Space, and Environmental Medicine* 62(3): 233–235, 1991.
Bohnen, H.G.M., and Gaillard, A.W.K. The effects of sleep loss in a combined tracking and time estimation task. *Ergonomics* 37(6): 1021–1030, 1994.
Borghouts, J., Soboczenski, F., Cairns, P., and Brumbly, D.P. Visualizing magnitude: Graphical number representations help users detect large number errors. Proceedings of the Human Factors and Ergonomics Society 59th Annual Meeting, 591–595, 2015.
Brand, J.L., and Judd, K.W. Angle of hard copy and text-editing performance. *Human Factors* 35(1): 57–70, 1993.
Briggs, R.W., and Goldberg, J.H. Battlefield recognition of armored vehicles. *Human Factors* 37(3): 596–610, 1995.
Brill, J.C., Rupert, A.H., and Lawson, B.D. Error analysis for localizing egocentric multimodal cues in the presence of helicopter noise. Proceedings of the Human Factors and Ergonomics Society 59th Annual Meeting, 1297–1301, 2015.
Casali, S.P., Williges, B.H., and Dryden, R.D. Effects of recognition accuracy and vocabulary size of a speech recognition system on task performance and user acceptance. *Human Factors* 32(2): 183–196, 1990.
Casner, S.M. Perceived vs. measured effects of advanced cockpit systems on pilot workload and error: Are pilots' beliefs misaligned with reality? *Applied Ergonomics* 40: 448–456, 2009.
Chan, A.H.S., Tsang, S.N.H., and Ng, A.W.Y. Effects of line length, line spacing, and line number on proofreading performance and scrolling of Chinese text. *Human Factors* 56(3): 521–534, 2014.
Chan, H.S., and Courtney, A.J. Stimulus size scaling and foveal load as determinants of peripheral target detection. *Ergonomics* 41(10): 1433–1452, 1998.
Chapanis, A. Short-term memory for numbers. *Human Factors* 32(2): 123–137, 1990.
Chen, H., Chan, K., and Tsoi, K. Reading self-paced moving text on a computer display. *Human Factors* 30(3): 285–291, 1988.
Chen, H., and Tsoi, K. Factors affecting the readability of moving text on a computer display. *Human Factors* 30(1): 25–33, 1988.
Chen, J.Y.C., and Barnes, M.J. Supervisory control of multiple robots: Effects of imperfect automation and individual differences. *Human Factors* 54(2): 157–174, 2012.
Chong, J., and Triggs, T.J. Visual accommodation and target detection in the vicinity of a window post. *Human Factors* 31(1): 63–75, 1989.
Christensen, C.L., Gliner, J.A., Horvath, S.M., and Wagner, J.A. Effects of three kinds of hypoxias on vigilance performance. *Aviation, Space, and Environmental Medicine* 48(6): 491–496, 1977.
Cohen, D., Otakeno, S., Previc, F.H., and Ercoline, W.R. Effect of "inside-out" and "outside-in" attitude displays on off-axis tracking in pilots and nonpilots. *Aviation, Space, and Environmental Medicine* 72(3): 170–176, 2001.
Colquhoun, W.P. The effect of 'unwanted' signals in performance in a vigilance task. *Ergonomics* 4(1): 41–52, 1961.
Cook, M.B., Smallman, H.S., Lacson, F.C., and Manes, D.I. Situation displays for dynamic UAV replanning: Intuitions and performance for display formats. Proceedings of the Human Factors and Ergonomics Society 54th Annual Meeting, 492–496, 2010.

Coury, B.G., Boulette, M.D., and Smith, R.A. Effect of uncertainty and diagnosticity on classification of multidimensional data with integral and separable displays of system status. *Human Factors* 31(5): 551–569, 1989.
Craig, A., Davies, D.R., and Matthews, G. Diurnal variation, task characteristics, and vigilance performance. *Human Factors* 29(6): 675–684, 1987.
Culley, K.E., Madhavan, P., Heikens, R., and Brown, J. Effects of emotional priming on visual threat detection. Proceedings of the Human Factors and Ergonomics 55th Annual Meeting, 232–236, 2011.
Curtis, M.T., Maraj, C., Ritman, M., and Jentsch, F. Investigation of the impact of feedback on decision accuracy and reaction time in a perceptual training task. Proceedings of the Human Factors and Ergonomics Society 54th Annual Meeting, 1630–1634, 2010.
Dodd, S., Lancaster, J., Miranda, A., Grother, S., DeMers, B., and Rogers, B. Touch screens on the flight deck: The impact of touch target size, spacing, touch technology and turbulence on pilot performance. Proceedings of the Human Factors and Ergonomics Society 58th Annual Meeting, 6–10, 2014.
Doll, T.J., and Hanna, T.E. Enhanced detection with bimodal sonar displays. *Human Factors* 31(5): 539–550, 1989.
Donderi, D.C. Visual acuity, color vision, and visual search performance. *Human Factors* 36(1): 129–144, 1994.
Downing, J.V., and Saunders, M.S. The effects of panel arrangement and focus of attention on performance. *Human Factors* 29(5): 551–562, 1987.
Drinkwater, B.L. Speed and accuracy in decision responses of men and women pilots. *Ergonomics* 11(1): 61–67, 1968.
Eatchel, K.A., Kramer, H., and Drews, F. The effects of interruption context on task performance. Proceedings of the Human Factors and Ergonomics Society 56th Annual Meeting, 2118–2122, 2012.
Eberts, R. Internal models, tracking strategies, and dual-task performance. *Human Factors* 29(4): 407–420, 1987.
Elvers, G.C., Adapathya, R.S., Klauer, K.M., Kancler, D.E., and Dolan, N.J. Effects of task probability on integral and separable task performance. *Human Factors* 35(4): 629–637, 1993.
Enander, A. Effects of moderate cold on performance of psychomotor and cognitive tasks. *Ergonomics* 30(10): 1431–1445, 1987.
Finomore, V., Popik, D., Castle, C., and Dallman, R. Effects of a network-centric multimodal communication tool on a communication monitoring task. Proceedings of the Human Factors and Ergonomics Society 54th Annual Meeting, 2125–2129, 2010.
Fisk, A.D., and Hodge, K.A. Retention of trained performance in consistent mapping search after extended delay. *Human Factors* 34(2): 147–164, 1992.
Fleury, M., and Bard, C. Effects of types of physical activity on the performance of perceptual tasks in peripheral and central vision and coincident timing. *Ergonomics* 30(6): 945–958, 1987.
Frankish, C., and Noyes, J. Sources of human error in data entry tasks using speech input. *Human Factors* 32(6): 697–716, 1990.
Frazier, J.W., Repperger, D.N., Toth, D.N., and Skowronski, V.D. Human tracking performance changes during combined $+G_z$ and $\pm G_y$ stress. *Aviation, Space, and Environmental Medicine* 53(5): 435–439, 1982.

Funke, G.J., Warm, J.S., Baldwin, C.L., Garcia, A., Funke, M.E., Dillard, M.B., Finomore, V.S., Mathews, G., and Greenlee, E.T. The independence and interdependence of coacting observers in regard to performance efficiency, workload, and stress in a vigilance task. *Human Factors* 58(6): 915–926, 2016.

Galinsky, T.L., Warm, J.S., Dember, W.N., Weiler, E.M., and Scerbo, M.W. Sensory alternation and vigilance performance: The role of pathway inhibition. *Human Factors* 32(6): 717–728, 1990.

Gerard, M.J., and Martin, B.J. Post-effects of long-term hand vibration on visuo-manual performance in a tracking task. *Ergonomics* 42(2): 314–326, 1999.

Griffin, W.C., and Rockwell, T.H. Computer-aided testing of pilot response to critical in-flight events. *Human Factors* 26(5): 573–581, 1984.

Hancock, P.A., and Caird, J.K. Experimental evaluation of a model of mental workload. *Human Factors* 35(3): 413–419, 1993.

Harris, C.S., and Johnson, D.L. Effects of infrasound on cognitive performance. *Aviation, Space, and Environmental Medicine* 49(4): 582–586, 1978.

Harris, C.S., and Shoenberger, R.W. Combined effects of broadband noise and complex waveform vibration on cognitive performance. *Aviation, Space, and Environmental Medicine* 51(1): 1–5, 1980.

Helton, W.S., and Head, J. Earthquakes on the mind: Implications of disasters for human performance. *Human Factors* 54(2): 189–194, 2012.

Hubbard, D.C. Inadequacy of root mean square as a performance measure. Proceedings of the Fourth International Symposium on Aviation Psychology, 698–704, 1987.

Imbeau, D., Wierwille, W.W., Wolf, L.D., and Chun, G.A. Effects of instrument panel luminance and chromaticity on reading performance and preference in simulated driving. *Human Factors* 31(2): 147–160, 1989.

Kennedy, R.S., Odenheimer, R.C., Baltzley, D.R., Dunlap, W.P., and Wood, C.D. Differential effects of scopolamine and amphetamine on microcomputer-based performance tests. *Aviation, Space, and Environmental Medicine* 61(7): 615–621, 1990.

Kidwell, B., Calhoun, G.L., Ruff, H.A., and Parasuraman, R. Adaptable and adaptive automation for supervisory control of multiple autonomous vehicles. Proceedings of the Human Factors and Ergonomics Society 56th Annual Meeting, 428–432, 2012.

Kimchi, R., Gopher, D., Rubin, Y., and Raij, D. Performance under dichoptic versus binocular viewing conditions: Effects of attention and task requirements. *Human Factors* 35(1): 35–56, 1993.

Kirschenbaum, S.S., Trafton, J.G., Schunn, C.D., and Trickett, S.B. Visualizing uncertainty: The impact on performance. *Human Factors* 56(3): 509–520, 2014.

Kopardekar, P., and Mital, A. The effect of different work-rest schedules on fatigue and performance of a simulated directory assistance operator's task. *Ergonomics* 37(10): 1697–1707, 1994.

Kuller, R., and Laike, T. The impact of flicker from fluorescent lighting on well-being, performance and physiological arousal. *Ergonomics* 41(4): 433–447, 1998.

Kvalseth, T.O. Human performance comparisons between digital pursuit and compensatory control. *Ergonomics* 21(6): 419–425, 1978.

Lanzetta, T.M., Dember, W.N., Warm, J.S., and Berch, D.B. Effects of task type and stimulus heterogeneity on the event rate function in sustained attention. *Human Factors* 29(6): 625–633; 1987.

Lee, D., Jeong, C., and Chung, M.K. Effects of user age and zoomable user interfaces on information searching tasks in a map-type space. Proceedings of the Human Factors and Ergonomics Society 54th Annual Meeting, 571–575, 2010.

Lee, M.D., and Fisk, A.D. Disruption and maintenance of skilled visual search as a function of degree of consistency. *Human Factors* 35(2): 205–220, 1993.

Li, G., and Oliver, S. Design factors of in-vehicle map display for efficient road recognition. Proceedings of the 1st Human Performance, Situation Awareness and Automation: User-Centered Design for the New Millennium, October 15–19, 2000.

Loeb, M., Noonan, T.K., Ash, D.W., and Holding, D.H. Limitations of the cognitive vigilance decrement. *Human Factors* 29(6): 661–674, 1987.

Maddox, M.E., and Turpin, J.A. The effect of number ordering and orientation on marking speed and errors for mark-sensed labels. *Human Factors* 28(4): 401–405, 1986.

Manzey, D., Reichenbach, J., and Onnasch, L. Human performance consequences of automated decisions aids in states of fatigue. Proceedings of the Human Factors and Ergonomics Society 53rd Annual Meeting, 329–333, 2009.

Matthews, M.L., Lovasik, J.V., and Mertins, K. Visual performance and subjective discomfort in prolonged viewing of chromatic displays. *Human Factors* 31(3): 259–271, 1989.

McCann, P.H. The effects of ambient noise on vigilance performance. *Human Factors* 11(3): 251–256, 1969.

McClernon, C.K., Miller, J.C., and Christensen, J.C. Variance as a method for objectively assessing pilot performance. Proceedings of the Human Factors and Ergonomics Society 56th Annual Meeting, 85–89, 2012.

McDougald, B.R., and Wogalter, M.S. Increased comprehension of warning pictorials with color highlighting. Proceedings of the Human Factors and Ergonomics Society Annual Meeting, 1769–1772, 2011.

McLeod, R.W., and Griffin, M.J. Effects of duration and vibration on performance of a continuous manual control task. *Ergonomics* 36(6): 645–659, 1993.

Meister, D. *Human Factors Testing and Evaluation*. New York: Elsevier, 1986.

Mertens, H.W., and Collins, W.E. The effects of age, sleep deprivation, and altitude on complex performance. *Human Factors* 28(5): 541–551, 1986.

Meuter, R.F.I., and Lacherez, P.F. When and why threats go undetected: Impacts of event rate and shift length on threat detection accuracy during airport baggage screening. *Human Factors* 58(2): 218–228, 2016.

Millar, K., and Watkinson, N. Recognition of words presented during general anesthesia. *Ergonomics* 26(6): 585–594, 1983.

Moseley, M.J., and Griffin, M.J. Whole-body vibration and visual performance: An examination of spatial filtering and time-dependency. *Ergonomics* 30(4): 613–626, 1987.

Mullin, J., and Corcoran, D.W.J. Interaction of task amplitude with circadian variation in auditory vigilance performance. *Ergonomics* 20(2): 193–200, 1977.

O'Hanlon, J., Schmidt, A., and Baker, C.H. Sonar Doppler discrimination and the effect of a visual alertness indicator upon detection of auditory sonar signals in a sonar watch. *Human Factors* 7(2): 129–139, 1965.

Oron-Gilad, T., Parmet, Y., and Benor, D. Interfaces for dismounted soldiers: Examination of non-perfect visual and tactile alerts in a simulated hostile urban environment. Proceedings of the Human Factors and Ergonomics Society 59th Annual Meeting, 145–149, 2015.

Payne, D.G., and Lang, V.A. Visual monitoring with spatially versus temporally distributed displays. *Human Factors* 33(4): 443–458, 1991.

Pepler, R.D. Warmth and performance: An investigation in the tropics. *Ergonomics* 2(10): 63–88, 1958.

Pitrella, F.D., and Kruger, W. Design and validation of matching tests to form equal groups for tracking experiments. *Ergonomics* 26(9): 833–845, 1983.

Potter, J.J., and Singhose, W.E. Effects of input shaping on manual control of flexible and time-delayed systems. *Human Factors* 56(7): 1284–1295, 2014.

Raaijmaker, J.G.W., and Verduyn, W.W. Individual difference and the effects of an information aid in performance of a fault diagnosis task. *Ergonomics* 39(7): 966–979, 1996.

Ralston, J.V., Pisoni, D.B., Lively, S.E., Greene, G.B., and Mullennix, J.W. Comprehension of synthetic speech produced by rule: Word monitoring and sentence-by-sentence listening tones. *Human Factors* 33(4): 471–491, 1991.

Reichenbach, J., Onnasch, L., and Manzey, D. Misuse of automation: The impact of system experience on complacency and automation bias in interaction with automated aids. Proceedings of the Human Factors and Ergonomics Society 54th Annual Meeting, 374–378, 2010.

Reinerman-Jones, L., Taylor, G., Sprouse, K., Barber, D., and Hudson, I. Adaptive automation as a task switching and task congruence challenge. Proceedings of the Human Factors and Ergonomics Society 55th Annual Meeting, 197–201, 2011.

Rhodes, W., Gugerty, L., Brooks, J., and Cantalupo, C. The effects of electronic map displays and spatial ability on performance of navigational tasks. Proceedings of the Human Factors and Ergonomics Society 53rd Annual Meeting, 369–373, 2009.

Rosenberg, D.J., and Martin, G. Human performance evaluation of digitizer pucks for computer input of spatial information. *Human Factors* 30(2): 231–235, 1988.

Ruffell-Smith, H.P. *A Simulator Study of the Interaction of Pilot Workload with Errors, Vigilance, and Decisions (TM 78432)*. Moffett Field, CA: NASA Ames Research Center, January 1979.

Salame, P., and Baddeley, A. Noise, unattended speech and short-term memory. *Ergonomics* 30(8): 1185–1194, 1987.

Schmidt, T.N., Teo, G.W.L., Szalma, J.L., Hancock, G.M., and Hancock, P.A. The effect of video game play on performance in a vigilance task. Proceedings of the Human Factors and Ergonomics Society 56th Annual Meeting, 1544–1547, 2012.

Siah, K.T., Yang, X.J., Yoshida, N., Ogiso, K., Holtta-Otto, K., and Naito, Y. Effects of experience, withdrawal speed and monitor size on colonoscopists' visual detection of polyps. Proceedings of the Human Factors and Ergonomics Society 59th Annual Meeting, 471–475, 2015.

Smith, A.P., and Miles, C. The effects of lunch on cognitive vigilance tasks. *Ergonomics* 29(10): 1251–1261, 1986.

St. John, M., and Risser, M.R. Sustaining vigilance by activating a secondary task when inattention is detected. Proceedings of the Human Factors and Ergonomics Society 53rd Annual Meeting, 155–159, 2009.

Swanson, L., Jones, E., Riordan, B., Bruni, S., Schurr, N., Sullivan, S., and Lansey, J. Exploring human error in an RPA target detection task. Proceedings of the Human Factors and Ergonomics Society 56th Annual Meeting, 328–332, 2012.

Taylor, G.S., and Szalma, J.L. The effects of the adaptability and reliability of automation on performance, stress and workload. Proceedings of the Human Factors and Ergonomics Society 53rd Annual Meeting, 160–164, 2009.

Teo, G.W., Schmidt, T.N., Szalma, J.L., Hancock, G.M., and Hancock, P.A. The effects of feedback in vigilance training on performance, workload, stress and coping. Proceedings of the Human Factors and Ergonomics Society 57th Annual Meeting, 1119–1123, 2013.

Teo, G., Schmidt, T., Szalma, J., Hancock, G.M., and Hancock, P.A. The effects of individual differences on vigilance training and performance in a dynamic vigilance task. Proceedings of the Human Factors and Ergonomics Society 58th Annual Meeting, 964–968, 2014.

Teo, G.W.L., and Szalma, J.L. The effect of spatial and temporal task characteristics on performance, workload, and stress. Proceedings of the Human Factors and Ergonomics Society 54th Annual Meeting, 1699–1703, 2010.

Teo, G., and Szalma, J.L. The effects of task type and source complexity on vigilance performance, workload, and stress. Proceedings of the Human Factors and Ergonomics Society 55th Annual Meeting, 1180–1184, 2011.

Thomas, E., Deleon, R., Kelling, N., and Harper, C. Arrow key configurations on laptop keyboards: Performance and user preference of the inverted-T and modified-T layout. Proceedings of the Human Factors and Ergonomics Society 59th Annual Meeting, 1071–1074, 2015.

Torle, G. Tracking performance under random acceleration: Effects of control dynamics. *Ergonomics* 8(4): 481–486, 1965.

Tullis, T.S. Human performance evaluation of graphic and textual CRT displays of diagnostic data. Proceedings of the Human Factors Society 24th Annual Meeting, 310–316, 1980.

Tzelgov, J., Henik, A., Dinstein, I., and Rabany, J. Performance consequences of two types of stereo picture compression. *Human Factors* 32(2): 173–182, 1990.

Van der Kleij, R., and te Brake, G. Map-mediated dialogues: Effects of map orientation differences and shared reference points on map location-finding speed and accuracy. *Human Factors* 52(4): 526–536, 2010.

Van Orden, K.F., Benoit, S.L., and Osga, G.A. Effects of cold air stress on the performance of a command and control task. *Human Factors* 38(1): 130–141, 1996.

Vermeulen, J. Effects of functionally or topographically presented process schemes on operator performance. *Human Factors* 29(4): 383–394, 1987.

Vidulich, M.A. The Bedford Scale: Does it measure spare capacity. Proceedings of the 6th International Symposium on Aviation Psychology, vol. 2, 1136–1141, 1991.

Vuckovic, A., Sanderson, P.M., Neal, A., Gaukrodger, S., and Wong, B.L.W. Relative position vectors: An alternative approach to conflict detection in air traffic control. *Human Factors* 55(5): 946–964, 2013.

Walker, B.N., Lindsay, J., Nance, A., Nakano, Y., Palladino, D.K., Dingler, T., and Jeon, M. Spearcons (speech-based earcons) improve navigation performance in advanced auditory menus. *Human Factors* 55(1): 157–182, 2013.

Wang, Y., Millet, B., and Smith, J.L. Informing the use of vibrotactile feedback for information communication: An analysis of user performance across different vibrotactile designs. Proceedings of the Human Factors and Ergonomics Society 58th Annual Meeting, 1859–1863, 2014.

Warren, W.R., Blickensderfer, B., Cruit, J., and Boquet, A. Shift turnover strategy and time in aviation maintenance. Proceedings of the Human Factors and Ergonomics Society 57th Annual Meeting, 46–50, 2013.

Wierwille, W.W., Rahimi, M., and Casali, J.G. Evaluation of 16 measures of mental workload using a simulated flight task emphasizing mediational activity. *Human Factors* 27(5): 489–502, 1985.
Wild-Wall, N., Hahn, M., and Falkenstein, M. Preparatory processes and compensatory effort in older and younger participants in a driving-like dual task. *Human Factors* 53(2): 91–102, 2011.
Yeh, Y., and Silverstein, L.D. Spatial judgments with monoscopic and stereoscopic presentation of perspective displays. *Human Factors* 34(5): 583–600, 1992.
Zaitzeff, L.P. Aircrew task loading in the Boeing multimission simulator. Proceedings of Measurement of Aircrew Performance – The Flight Deck Workload and its Relation to Pilot Performance (AGARD-CP-56). AGARD, Neuilly-sur-Seine, France, 1969.
Zeiner, A.R., and Brecher, G.A. Reaction time with and without backscatter from intense pulsed light. *Aviation, Space, and Environmental Medicine* 46(2): 125–127, 1975.

2.2 Time

The second category of human performance measures is time. Measures in this category assume that tasks have a well-defined beginning and end so that the duration of task performance can be measured. Measures in this category include dichotic listening detection time (Section 2.2.1), glance duration (Section 2.2.2), lookpoint time (Section 2.2.3), marking speed (Section 2.2.4), movement time (Section 2.2.5), reaction time (Section 2.2.6), reading speed (Section 2.2.7), search time (Section 2.2.8), task load (Section 2.2.9), and time to complete (Section 2.2.10).

2.2.1 Dichotic Listening Detection Time

General description – Participants are presented with auditory messages through headphones. Each message contains alphanumeric words. Different messages are presented simultaneously to each ear. Participants must detect specific messages when they are presented in a specific ear.

Strengths and limitations – Gopher (1982) reported significant differences in scores between flight cadets who completed training and those who failed.

Data requirements – Auditory recordings are required as well as an audio system capable of displaying different messages to each ear. Finally, a system is required to record omissions, intrusions, and switching errors.

Thresholds – Low threshold is zero. High threshold was not stated.

Source

Gopher, D. Selective attention test as a predictor of success in flight training. *Human Factors* 24(2): 173–183, 1982.

2.2.2 Glance Duration

General description – The duration that a human visually samples a single scene is a glance. Glance duration has been used to evaluate controls, displays, and procedures.

Strengths and limitations – Glance duration has long been used to evaluate driver performance. In an early study, Mourant and Rockwell (1970) analyzed the glance behavior of eight drivers traveling at 50 mph (80.47 kph) on an expressway. As the route became more familiar, drivers increased glances to the right edge marker and horizon. While following a car, drivers glanced more often at lane markers. Imbeau et al. (1989) used time glancing at a display to evaluate instrument panel lighting in automobiles. Not unexpectedly, higher complexity of messages was associated with significantly longer (+0.05 seconds more) glance times. Similarly, Land (1993) reported that drivers' glance behavior varies throughout a curve: a search for cues during approach, into the curve during the bend, and target beyond the bend on exit.

Mourant and Donohue (1977) reported that novice and young experienced drivers made fewer glances to the left outside mirror than did mature drivers. Novice drivers also made more direct looks than glances in the mirrors prior to executing a maneuver. In addition, novice drivers have longer and more frequent glances at vehicles than at obstacles (Masuda et al., 1990). Glance duration increases as blood alcohol concentration level increases (Masuda et al., 1990).

There has been some nondriver research as well. For example, Fukuda (1992) reported a maximum of 6 characters could be recognized in a single glance.

Data Requirements – Eye movements must be recorded to an accuracy of ±0.5 degrees horizontal and ±1 degree vertical (Mourant and Rockwell, 1970).

Thresholds – Minimum glance duration 0.68 s (Mourant and Donohue, 1977). Maximum glance duration 1.17 s (Mourant and Donohue, 1977).

Sources

Fukuda, T. Visual capability to receive character information Part I: How many characters can we recognize at a glance? *Ergonomics* 35(5): 617–627, 1992.

Imbeau, D., Wierwille, W.W., Wolf, L.D., and Chun, G.A. Effects of instrument panel luminance and chromaticity on reading performance and preference in simulated driving. *Human Factors* 31(2): 147–160, 1989.

Land, M.F. Eye-head coordination during driving. IEEE Systems, Man and Cybernetics Conference Proceedings, 490–494, 1993.

Masuda, K., Nagata, M., Kureyama, H., and Sato, T.B. *Visual Behavior of Novice Drivers as Affected by Traffic Conflicts (SAE Paper 900141)*. Warrendale, PA: Society of Automotive Engineers, 1990.
Mourant, R.R., and Donohue, R.J. Acquisition of indirect vision information by novice, experienced, and mature drivers. *Journal of Safety Research* 9(1): 39–46, 1977.
Mourant, R.R., and Rockwell, T.H. Mapping eye movement patterns to the visual scene in driving: An exploratory study. *Human Factors* 12(1): 81–87, 1970.

2.2.3 Lookpoint Time

General description – Lookpoint is "the current coordinates of where the pilot is looking during any one thirtieth of a second" (Harris et al., 1986, p. 38). Lookpoint is usually analyzed by either real-time viewing of lookpoint superimposed on the instrument panel or examination of time histories.

Strengths and limitations – Real-time observation of lookpoint efficiently informs the researcher about scanning behavior as well as helps to identify any calibration problems. Analysis of time histories of lookpoint provides information such as average dwell time, dwell percentage, dwell time, fixation, fixations per dwell, one- and two-way transitions, saccades, scans, transition, and transition rate. Such information is useful in (1) arranging instruments for optimum scanning, (2) assessing the time required to assimilate information from each display, and (3) estimating the visual workload associated with each display and task criticality (i.e., blink rate decreases as task criticality increases) (Stern et al., 1984). Corkindale (1974) reported significant differences between working conditions in the percentage of time looking at a Head Up Display (HUD). Spady (1977) reported different scanning behavior during approach between manual (73% of time on flight director, 13% on airspeed) and autopilot with manual throttle (50% on flight director, 13% on airspeed).

Limitations of lookpoint include: (1) inclusion of an oculometer into the workplace or simulator, (2) requirement for complex data-analysis software, (3) lack of consistency, (4) difficulty in interpreting the results, and (5) lack of sensitivity of average dwell time. The first limitation is being overcome by the development of miniaturized oculometers, the second by the availability of standardized software packages, the third by collecting enough data to establish a trend, the fourth by development of advanced analysis techniques, and the fifth by use of the dwell histogram.

Data requirements – The oculometer must be calibrated and its data continuously monitored to ensure that it is not out of track. Specialized data reduction and analysis software is required.

Thresholds – Not stated.

Sources

Corkindale, K.G.G. A flight simulator study of missile control performance as a function of concurrent workload. Proceedings of Simulation and Study of High Workload (AGARD-CP-146), 1974.

Harris, R.L., Glover, B.J., and Spady, A.A. *Analytical Techniques of Pilot Scanning Behavior and Their Application (NASA Technical Paper 2525).* Hampton, VA: NASA Langley, July 1986.

Spady, A.A. Airline pilot scanning behavior during approaches and landing in a Boeing 737 simulator. Proceedings of Guidance and Control Design Considerations for Low Altitude and Terminal Area Flight (AGARD-CP-240), 1977.

Stern, J.A., Walrath, L.C., and Goldstein, R. The indigenous eye blink. *Psychophysiology* 21(1): 22–23, 1984.

2.2.4 Marking Speed

General description – Maddox and Turpin (1986) used speed to evaluate performance using mark-sensed forms.

Strengths and limitations – Marking speed was not affected by number orientation (horizontal or vertical), number ordering (bottom-to-top or top-to-bottom), or handedness of users (Maddox and Turpin, 1986).

Data requirements – Start and stop times must be recorded.

Thresholds – The average marking speed was 4.74 seconds for a five-digit number.

Source

Maddox, M.E., and Turpin, J.A. The effect of number ordering and orientation on marking speed and errors for mark-sensed labels. *Human Factors* 28(4): 401–405, 1986.

2.2.5 Movement Time

General description – Arnaut and Greenstein (1990) provided three definitions of movement time. "Gross movement was defined as the time from the initial touch on the [control] … to when the cursor first entered the target. Fine adjustment was the time from the initial target entry to the final lift-off of the finger from the [control]. … Total movement time was the sum of these two measures" (p. 655).

Strengths and limitations – Arnaut and Greenstein (1990) reported significant increases in gross movements and significant decreases in fine adjustment times for larger (120 mm) than smaller (40 mm) touch tablets. Total movement times were significantly longer for the largest and smallest touch tablets than for the intermediate size tablets (60, 80, or 100 mm). For a trackball, gross and total movement times were significantly longer for a longer distance (160 or 200 mm) than the shortest distance (40 mm). In a second experiment, there were no significant differences in any of the movement times for the touch tablet with and without a stylus. All three measures were significantly different between different display amplitudes and target widths. "In general performance was better with the smaller display amplitudes and the larger display target widths" (p. 660).

Lin et al. (1992) used movement time to assess a head-controlled computer input device. They reported the longest movement time for small width targets (2.9 mm versus 8.1 or 23.5 mm), high gains (1.2 versus 0.15, 0.3, or 0.6), and large movement amplitudes (61.7 mm versus 24.3 mm). Hancock and Caird (1993) reported that movement time increased as the shrink rate of a target increased. Further movement time decreased as path length increased.

Hoffman and Sheikh (1994) reported movement times increased as target height decreased from 200 to 1 mm and also as target width decreased from 40 to 10 mm. Li et al. (1995) reported that movement time increased with distance moved but disproportionately increased at extreme height (1,080 mm above seat reference point) and angle (90 degrees ipsilaterally).

Rorie et al. (2013) used movement time to compare differences in target size, target distance, spring force level, and gravitational force level using a Novint Falcon input device. There were significant differences in time for target size, target distance (farther was longer), and gravitational force level (higher was faster) but not for spring force level. Sheik-Nainar et al. (2013) reported no difference in movement time in a comparison of force input (light versus heavy press) on touch pads. In a similar study, Kar et al. (2015) reported significantly slower movement time for a 3D motion-and-gesture control interface than with a mouse or trackpad.

Data requirements – The time at both the beginning and the end of the movement must be recorded.

Thresholds – Movement times below 50 msec are very uncommon.

Sources

Arnaut, L.Y., and Greenstein, J.S. Is display/control gain a useful metric for optimizing an interface? *Human Factors* 32(6): 651–663, 1990.

Hancock, P.A., and Caird, J.K. Experimental evaluation of a model of mental workload. *Human Factors* 35(3): 413–419, 1993.

Hoffman, E.R., and Sheikh, I.H. Effect of varying target height in a Fitts' movement task. *Ergonomics* 36(7): 1071–1088, 1994.

Kar, G., Vu, A., Nehme, B.J., and Hedge, A. Effects of mouse, trackpad, and 3D motion and gesture control on performance, posture, and comfort. Proceedings of the Human Factors and Ergonomics Society 59th Annual Meeting, 327–331, 2015.
Li, S., Zhu, Z., and Adams, A.S. An exploratory study of arm-reach reaction time and eye-hand coordination. *Ergonomics* 38(4): 637–650, 1995.
Lin, M.L., Radwin, R.G., and Vanderheiden, G.C. Gain effects on performance using a head-controlled computer input device. *Ergonomics* 35(2): 159–175, 1992.
Rorie, R.C., Vu, K.L., Marayong, P., Robles, J., Strybel, T.Z., and Battiste, V. Effects of type and strength of force feedback on movement time in a target selection task. Proceedings of the Human Factors and Ergonomics Society 57th Annual Meeting, 36–40, 2013.
Sheik-Nainar, M., Ostberg, A., and Matic, N. Two-level force input on touchpad and the effects of feedback on performance. Proceedings of the Human Factors and Ergonomics Society 57th Annual Meeting, 1052–1056, 2013.

2.2.6 Reaction Time

General description – RT is the time elapsed between stimulus onset and response onset. The stimulus is usually a visually presented number requiring a manual key press but any stimulus and any input or output mode are possible.

Strengths and limitations – RT may measure the duration of mental processing stages (Donders, 1969). RT is sensitive to physiological state such as fatigue, sleep deprivation, aging, brain damage, and drugs (Boer et al., 1984; Frowein, 1981; Frowein et al., 1981a, 1981b; Gaillard et al., 1983, 1985, 1986; Logsdon et al., 1984; Moraal, 1982; Sanders et al., 1982; Steyvers, 1987).

Fowler et al. (1987) used RT for the first correct response to a five-choice visual RT task to examine the effects of hypoxia. The response time was greater at 82% arterial oxyhemoglobin saturation than at 84 or 86%. In a related study, Fowler et al. (1989) reported increased RT as a function of inert gas narcosis.

RT is also sensitive to the effects of time. For example, Coury et al. (1989) reported significant decreases in RT over time. Harris et al. (1995) reported significant decreases in response time as time on task increased. Pigeau et al. (1995) measured response time of air defense operators to detect incoming aircraft. There was a significant interaction of shift and time on task. Participants working the midnight shift had longer RTs in the 60/60-minute work/rest schedule than those in the evening shift. There was also a shift-by-zone significant interaction: longer RTs for midnight shift for northern regions (i.e., low air traffic). Finally, between the two sessions, RT increased during the midnight shift but decreased during the evening shift.

RT is a reliable measure (e.g., split-half reliabilities varying between 0.81 and 0.92, AGARD, 1989, p. 12). However, Vidulich (1991) reported a test-retest reliability of +0.39 of a visual choice RT task. However, Boer (1987) suggested that RT performance may require 2,000 trials to stabilize. Further, Carter et al. (1986) reported that slope was less reliable than RT for a choice RT task.

RT has been used to evaluate auditory, tactile, visual, and vestibular stimuli. An early study compared all three. Swink (1966) compared RTs to a light, a buzzer, and an electro-pulse. RTs were shortest for the pulse.

2.2.6.1 Auditory Stimuli

Early work in the 1960s examined the effects of stimulus characteristics on RT. For example, Loeb and Schmidt (1963) measured the RT of eight participants to randomly occurring auditory signals over eight 50-minute sessions. In four sessions the signal was 10 dB above threshold. In the other four it was 60 dB above threshold. RT in the 10 dB sessions increased if the responses were merely acknowledged rather than the participants being told the signal would be faster or slower. In a six-month study, Warrick et al. (1965) measured the RT of five secretaries to a buzzer which was activated without warning once or twice a week. RT decreased over time and with alerting. Simon (1967) reported significantly longer RTs with cross stimulus response correspondence (i.e., responding to an auditory stimulus in the left ear with the right hand) than for same side response. This was especially true for older participants.

Later work focused on communication and feedback. For example, Payne and Lang (1991) reported shorter RTs for rapid communication than for conventional visual displays. In another communication experiment, RTs were significantly faster for target words in natural than in synthetic speech (Ralston et al., 1991). Akamatsu et al. (1995) reported that there were no differences in RT associated with the type of feedback (normal, auditory, color, tactile, and combined) provided by a computer mouse.

Begault and Pittman (1996) used detection time to compare conventional versus 3-D audio warnings of aircraft traffic. Detection time was significantly shorter (500 msec) for the 3-D audio display. Haas and Casali (1995) reported that RT to an auditory stimulus decreased as perceived urgency increased and as pulse level increased. Donmez et al. (2009) reported that course deviation sonifications resulted in significantly faster RTs than a discrete alert for military operators of Unmanned Aerial Vehicles (UAVs). Robinson et al. (2012) reported significantly faster time to locate a simulated sniper with 3-D audio cues than with either a semantic cue or a monaural cue.

Like other RTs to other stimuli, Horne and Gibbons (1991) reported a significant increase in RT with increases in alcohol dose. In a personality trait study, Dattel et al. (2011) reported that participants higher in inattentional blindness took 50% longer to answer irrelevant questions about a video driving scene than those lower in inattentional blindness. In a more recent study, Arrabito et al. (2013) reported significantly faster RT auditory display than tactile display warnings.

2.2.6.2 Tactile Stimuli

In moving-base simulators, RT to crosswind disturbances was significantly shorter when physical-motion cues were present than when they were not present (Wierwille et al., 1983). Chancey et al. (2014) used RT to compare vibrotactile stimuli of varying wave forms, inter-pulse intervals, and envelopes.

2.2.6.3 Visual Stimuli

The vast majority of the research using RT has been performed with visual stimuli. The research has examined the effects of task, environment, and participant variables. There have also been efforts to break RT into components and models.

Task Variables Task variables examined using RT include target characteristics, display formats, task scheduling, and task type.

Target characteristics In an early study, Baker et al. (1960) reported increases in RT as the number of irrelevant items increased and as the difference in resolution of reference and form increased. RT increased as stimulus complexity (1, 2, or 4 vertical lines) increased and inter-stimulus interval (100, 300, 500, 700, 900, and 1,100 ms) decreased (Aykin et al., 1986). Mackie and Wylie (1994) reported decreased RT to sonar signals with performance feedback and increased signal rate. Hancock and Caird (1993) reported a significant decrease in RT as the length of a path from the cursor to the target increased. Teo and Szalma (2010) reported significant differences in RT on correct trials with the fastest responses made in the 1-display condition and the slowest in the 8-display condition.

In a field study, Cole et al. (1978) recorded the RTs of Air Traffic Controllers (ATC) to aircraft initial movement prior to takeoff. The independent variables were delay in takeoff, angle of observer's line of sight to the centerline of the runway, experience of the ATC personnel, speed of aircraft acceleration, and use of binoculars. RT was significantly related to each of these independent variables.

Yeh and Silverstein (1992) measured RT as participants made spatial judgments to a simplified aircraft landing display. They reported significantly shorter RT for targets in the high front versus low back portions of the visual field of the display, for larger altitude separations between targets, for the 45-degree versus the 15-degree viewing orientation, and with the addition of binocular disparity. Steelman and McCarley (2011) reported significantly shorter RTs when target expectancy was higher. RTs were also shorter when the targets were near the center of the display.

Display formats investigated include flashing, background, color, viewing angle, and type of symbology.

In an early study on flashing, Crawford (1962) reported that RTs were longer for a visual stimulus against a background of flashing lights than against a background of steady lights. However, Wolcott et al. (1979) found no significant differences in choice RT to light flashes.

The effect of background has often been investigated. Thackray et al. (1979) reported significant increases in RT to critical targets in the presence of 16 targets rather than four or eight targets. Zeiner and Brecher (1975) reported significant increases in RTs to visual stimuli in the presence of backscatter from strobe lights. Hollands and Lamb (2011) reported significantly shorter RTs for exocentric than for tethered displays but no difference between these and egocentric displays.

Lee and Fisk (1993) reported faster RTs in a visual search task if the consistency of the stimuli remained 100% than if it did not (67, 50, or 33% consistent). Holahan et al. (1978) asked participants to state "stop" or "go" when presented a photo of a traffic scene with or without a stop sign. RTs increased as the number of distracters increased and in the presence of some red in the distracters. In another traffic study, Ho et al. (2001) measured RT to traffic signs in visual scenes with varying amounts of clutter. There were significant effects of age (younger drivers had shorter RTs), clutter (clutter increased RTs), and presence of target (presence made RTs shorter).

Related to background are the effects of redundant cues. Simon and Overmeyer (1984) used RT to evaluate the effects of redundant visual cues. Simon et al. (1988) in contrast examined the effects of redundant versus different cues on RT. Jubis (1990) used RT to evaluate display codes. She reported significantly faster RT for redundant color and shape (mean = 2.5 s) and color (mean = 2.3 s) coding than for partially redundant color (mean = 2.8 s) or shape (mean = 3.5 s). Perrott et al. (1991) reported significant decreases in RT to a visual target when spatially correlated sounds were presented with the visual targets. Also, MacDonald and Cole (1988) reported a significant decrease in RT for flight tasks in which relevant information was color coded as compared to monochromatic displays.

Murray and Caldwell (1996) reported significantly longer RTs as the number (1, 2, 3) of displays to be monitored increased and number (1, 2, 3) of display figures increased. There were also longer RTs to process independent rather than redundant images. Hess et al. (2013) compared targets of various sizes against a camouflaged background and reported significantly longer RTs as target size decreased.

Tzelgov et al. (1990) used RT to compare two types of stereo picture compression. They reported significant task (faster RTs for object decision than depth decision task), depth (faster RTs for smaller depth differences), size (faster RTs for no size difference between compared objects), and presentation effects. There were also numerous interactions.

Tullis (1980) reported significant difference on RT as a function of display format. Color graphics were associated with significantly shorter RTs than narrative text with structured text and black and white graphics in between.

In an early study on viewing angle, Simon and Wolf (1963) reported increases in RT as viewing angle increased. Related to viewing angle is layout. Downing and Sanders (1987) reported longer RTs in simulated control room emergencies with mirror image panels than non-mirror image panels.

The effects of type of symbology have been investigated in aviation using RT. For example, Remington and Williams (1986) used RT to measure the efficiency with which helicopter situation display symbols could be located and identified. RTs for numeric symbols were significantly shorter than for graphic symbols. Negative trials (target not present) averaged 120 ms longer than positive trials (target present), however, there were more errors on positive than on negative trials.

Taylor and Selcon (1990) compared RT for four display formats (HUD, Attitude Indicator, Aircraft Reference, and Command Indicator) in a simulated aircraft unusual attitude recovery. There were significantly longer RTs for the Aircraft Reference display. In another aviation experiment, Moacdieh et al. (2013) reported a significant increase in RT to visual alerts as the amount of clutter in the Primary Flight Display (PFD) increased. In a more recent study, Moacdieh and Sarter (2015) reported high data density and poor organization of visually presented information resulted in significantly longer RTs. Pigeau et al. (1995) measured RT of air defense operators to detect incoming aircraft. Significantly longer RTs occurred when the same geographical area was displayed in two rather than four zones.

Symbology has also been investigated in other domains. For example, Chapanis and Lindenbaum (1959) used RT to stove control-burner arrangements. They found a decrease in RT over trials 1 through 40 but not over trials 41 through 81. They also found significantly shorter RTs with one of the four stove control-burner configurations. Buttigieg and Sanderson (1991) reported significant differences in RT between display formats. There were also significant decreases in RT over the three days of the experiment.

Coury et al. (1989) reported significantly faster RTs to a digital display than to a configural or bar graph display. The task was classification. In a similar display study, Steiner and Camacho (1989) reported significant increases in RT as the number of bits of information increased especially for alphanumerics as compared to icons. In addition, Imbeau et al. (1989) used the time from stimulus presentation to a correct answer to evaluate driver reading performance. These authors reported significantly longer RTs for 7-arcmin characters (3.45 to 4.86 s) than for 25-arcmin characters (1.35 s). In a map-viewing task, Rizzardo and Colle (2013) reported significantly faster RT to left/right turn driving decisions when spatial with verbal turn indicators were provided rather than spatial only indicators.

In another driving study, Kline et al. (1990) converted visibility distances of road signs to sight times in seconds assuming a constant travel speed. This measure made differences between icon and text signs very evident. Also, McKnight and Shinar (1992) reported significantly shorter brake RT in following vehicles when the forward vehicle had center high-mounted stop lamps.

RT has also been useful in discriminating display utility (Nataupsky and Crittenden, 1988). Further, Boehm-Davis et al. (1989) reported faster RT to database queries if the format of the database was compatible with the information sought. Perriment (1969) examined RT to bisensory stimuli, i.e., auditory and visual, and Morgerstern and Haskell (1971) reported significantly slower RTs for visual than for auditory stimuli. Finally, Curtis et al. (2010) reported significant increases in RT from pretest to training and then a decrease in the post test for a visual discrimination task when either no feedback or conceptual feedback was given to participants. When knowledge of results was given, however, RT consistently decreased across all three time periods.

Task scheduling Adams et al. (1962) reported a significant decrease in RT with a three-hour visual detection task but not over nine days of task performance. In the same year, Teichner (1962) reported RT increased as the probability of detection during an initial session increased.

Task type Dewar et al. (1976) reported shorter RTs for classifying than identifying visually presented traffic signs. RTs were also shorter for warning than for regulatory signs and for text than symbol signs. Wierwille et al. (1985) reported that RT was significantly affected by the difficulty of a mathematical problem-solving task. They defined RT as the time from problem presentation to a correct response.

Koelega et al. (1989) reported no significant differences in RT between four types of visual vigilance tasks: "(1) the cognitive Continuous Performance Test (CPT), in which the target was the sequence of the letters AX; (2) a visual version of the cognitive Bakan task in which a target was defined as three successive odd but unequal digits; (3) an analogue of the Bakan using non-digital stimuli; and (4) a pure sensory task in which the critical signal was a change in brightness" (p. 46).

Fisk and Jones (1992) reported significant effects of search consistency (varying the ratio of consistent [all words were targets] to inconsistent words: 8:0, 6:2, 4:4, 2:6, and 0:8; shorter RTs with increased consistency) and practice (shorter RT from trial 1 to 12) on correct RT. Elvers et al. (1993) reported significant decreases in RT as a function of practice. This was especially true for a volume estimation versus a distance estimation task.

Kimchi et al. (1993) reported significantly shorter RTs for a local-directed rather than a global-directed task in a focused than in a divided attention condition. Significantly shorter RTs were reported for the global-directed than the local-directed task but only in the divided attention condition. Van Assen and Eijkman (1971) used RT to assess learning in a two-dimensional tracking task.

There have even been studies to assess the effect of how a task is performed. For example, Welford (1971) reported RTs were longer when made by the ring and middle fingers than when made by the index and little fingers.

<u>Environment Variables</u> In an early environment study, Chiles (1958) reported no significant differences associated with temperature. However,

Shvartz et al. (1976) reported 30% longer RTs during exercise in temperate (23 degrees Centigrade) and hot environments (30 degrees Centigrade) than during rest. This reduced to 10% higher RT after eight days of heat acclimation. Liebowitz et al. (1972) reported no significant difference in RT to red lights in the central visual area while participants were on a treadmill in a heat chamber with or without fluid replacement. There was a decrease in RT to lights in the periphery, however, with practice. But Enander (1987) reported no increase in RT in moderate cold (+5 degrees Centigrade).

Miles et al. (1984) reported longer RTs in noise environments (95 dBC) versus quiet (65 dBC). Warner and Heimstra (1972) reported a significant difference in RT as a function of noise (0, 80, 90, 100 dB), task difficulty (8, 16, 32 letter display), and target location (central and peripheral). Beh and Hirst (1999) reported shorter RTs to a stop-light display during music than during quiet. However, RT to peripheral visual targets was increased with high intensity (i.e., 85 or 87 dBA) music.

Macintosh et al. (1988) reported significantly longer RTs for those individuals showing effects of Acute Mountain Sickness at either 4,790 or 5,008 meters altitude. The RT task was a visual three-choice task. In another hypoxia experiment, Fowler and Kelso (1992) reported increased RT for hypoxia than normoxia especially for low intensity (0.38 cd/m^2) than high intensity (1.57 cd/m^2) visual stimuli (males and female names). Fowler et al. (1982) reported that hypoxia increased RT especially with lower luminance stimuli. Fowler et al. (1985) reported also increased RT with hypoxia. Leifflen et al. (1997) used the manikin choice RT task to evaluate the effects of hypoxia. There were no significant effects on either the number of errors or the RT due to hypoxia associated with even 7,000 m simulated altitude. However, Phillips et al. (2009) reported significant longer RTs associated with hypoxia.

McCarthy et al. (1995) reported significantly slower RTs in judging the orientation of visual targets at 7,000 and 12,000 feet relative to sea level. In a similar study, Linden et al. (1996) reported significant increases in RT associated with hypoxia and with increased angle of rotation of a target. Gliner et al. (1979) did not report any significant differences in detecting 1 second light pulses associated with ozone detection (0.00, 0.25, 0.50, and 0.75 ppm) when the rate of signals to nonsignals was low (1 out of 30). There was a decrease in performance at 0.75 parts per million (ppm) when the signal to non-signal ration was increased to 1 to 6, however. In an unusual study, Helton and Head (2012) reported changes in RT related to individual responses to the stress of experiencing a 7.1-magnitude earthquake. Heimstra et al. (1980) reported no effect of deprivation of cigarette smoking on RT.

Mertens and Collins (1986) used RT to warning light changes in visual pointer position and successive presentation of targets to evaluate the effects of age (30 to 39 versus 60 to 69 years old), altitude (ground versus 3,810 m), and sleep (permitted versus deprived). RT was standardized and then transformed so that better performance was associated with a higher score. Older persons had lower scores in the light task than younger persons. Sleep deprivation decreased performance on all three RT tasks; altitude did not significantly affect performance on either task. However, Alfred and Rice (2012) compared reaction time to visual stimuli of two groups of Army personnel aged 21 to 38 and aged 39 to 58. They reported no significant differences in RT.

In an aviation-related study (Stewart and Clark, 1975), 13 airline pilots were exposed to rotary accelerations (0.5, 1.0, and 5.0 degrees per second^2) prior (0, 50, 90, 542, 926, 5,503, and 9,304 ms) to performance of a visual choice RT task. RT increased significantly with increases in each independent variable.

As a caution when using RT, Krause (1982), based on data from 15 Navy enlisted men performing 50 trials on each of 15 days, recommended using at least 1,000 practice trials prior to using RT to assess environmental effects.

Participant Variables Age, gender, and experience have been investigated using RT.

Deupree and Simon (1963) reported longer RTs in the two-choice than in the single choice condition. Older participants (median age 75) moved faster in the simple than in the choice RT condition. Korteling (1990) compared visual choice RTs among diffuse brain-injury patients, older males (61 to 73 years old), and younger males (21 to 43 years old). The RTs of the patients and older males were significantly longer than the RTs of the younger males. RTs were significantly different as a function of response-stimulus interval (RSI; RSI = 100 msec, RT = 723 msec; 500, 698; 1,250, 713, respectively). There was also a significant RSI by stimulus sequence interaction, specifically, "interfering aftereffects of alternating stimuli decreased significantly with increasing RSI" (p. 99).

Adam et al. (1999) reported a trend for males to have shorter RTs than females. In a simulated ATC task, Thackray et al. (1978) reported a significant increase in RT over a two-hour session. There were, however, no significant differences between the 26 men and 26 women who participated in the experiment. Matthews and Ryan (1994) reported longer RTs to a line length comparison task in the premenstrual phase than in the intermenstrual phase. The participants were female undergraduate students.

DeMaio et al. (1978) asked instructor and student pilots to indicate if photos of an aircraft cockpit were correct. Instructors responded more quickly and more accurately. Briggs and Goldberg (1995) used RT to evaluate armored tank recognition ability. There were significant differences between participants, presentation time (shorter RTs as presentation time increased), view (flank view faster than frontal view), and model (M1 fastest, British Challenger slowest). There were no significant effects of component versus friend or foe.

RT is not always sensitive to differences among participants, however. For example, Leonard and Carpenter (1964) reported no correlation between a five-choice RT task and the A.H.4 Intelligence Task. There was, however, a significant correlation between RT and typewriter task performance measured as number of words typed in five minutes. Further, Park and Lee (1992) reported RT in a computer-aided aptitude task did not predict performance of flight trainees.

Swanson et al. (2012) used RT to evaluate Remotely Piloted Aircraft imagery exploitation. RTs were significantly shorter for low than high Ground Separation Distance, long than short dwell time, and long than short aspect angle. There was no significant effect of long versus short inter-event time.

Components Taking a different approach, Jeeves (1961) broke total RT into component times to assess the effects of advanced warning. Hilgendorf (1966) reported RT was a $\log_2 n$ function, with n being the number of bits per stimulus, and was linear even at high ns. Krantz et al. (1992) went one step farther and developed a mathematical model to predict RT as a function of spatial frequency, forward field of view/display luminance mismatch, and luminance contrast. Shattuck et al. (2013) reported no significant difference in RT as a function of exposure to waterborne motion.

In an unusual experiment, Helton et al. (2015) used RT in a vigilance task to compare levels of Dissociative Experiences Scale scores. In another unusual study, Claypoole and Szalma (2015) reported significantly shorter RTs in a vigilance task in the presence of a supervisor than when the supervisor was not present. Brown et al. (2015) reported 30% faster RTs when working on a three hours on/9 hours off schedule than on a five hours on/10 hours off schedule. In a more recent study, Jipp (2016) reported significantly slower RTs to consecutive automation failures than to the initial automation failures.

2.2.6.4 Vestibular Stimuli

Gundry (1978) asked participants to detect roll motion while seated on a turntable. There were no significant differences in RT for left versus right roll except when participants were given a visual cue. However, they detected rightwards roll significantly faster (236 ms) than leftwards roll (288 ms).

2.2.6.5 Related Measures

Related measures include detection time and recognition time. Detection time is defined as the onset of a target presentation until a correct detection is made. Recognition time is the onset of the target presentation until a target is correctly recognized (Norman and Ehrlich, 1986). In an early study, Haines (1968) reported a significant increase in detection time of a spot light source with the introduction of a star background (either real or simulated) or a glare source. His participants were 127 untrained observers in a planetarium. Tear et al. (2013) reported faster detection times with a Head-Mounted Display than with a monitor. Pascale et al. (2015) reported longer times to detect peripheral events when participants were wearing Google Glass™ rather than using a traditional computer monitor. Kerstholt et al. (1996) reported a significant increase in detection time with simultaneous targets.

There was also a significant increase in detection time for subsequent targets (first versus second versus third) in two complex target conditions.

Thackray and Touchstone (1991) used detection times to secondary targets in a simulated ATC task to examine the effect of color versus flashing as a cue under low and high task load. In the area of automation, Stelzer and Klein (2011) used the time for Air Traffic Controllers to detect a pilot deviation with and without automation support. There was no significant difference in time to detect although there had been a significance in the number detected. Further, the controllers found deviations significantly faster in low traffic loads than in high traffic loads.

In an early study, Johnston (1968) reported significant effects of horizontal resolution, shades of gray, and slant range on target recognition time. There were 12 participants with 20/20 vision or better. The targets were models of three military vehicles (2½ ton cargo truck, 5-ton flatbed truck, and tank with 90 mm gun) on a terrain board viewed through a closed-circuit television system.

Bemis et al. (1988) did not find any significant differences in threat detection times between conventional and perspective radar displays. However, response time to select the interceptor nearest to the threat was significantly shorter with the perspective display. Finomore et al. (2010) also used response time but to a communication task. They reported significant effects of difficulty (easy versus hard) and communication format (radio, 3D audio, chat, and multi modal).

Damos (1985) used a variant of RT, specifically, the average interval between correct responses (CRI). CRI includes the time to make incorrect responses. CRI was sensitive to variations in stimulus mode and trial. Finally, Wickens and Ward (2017) used percentage time that an aircraft was in a predicted conflict to compare traditional and 3D displays. Aircraft had significantly greater percentage time in conflict with the 3D displays than with the traditional 2D display.

Data requirements – Start time of the stimulus and response must be recorded to the nearest msec.

Thresholds – RTs below 50 msec are very uncommon.

Sources

Adam, J.J., Paas, F.G.W.C., Buekers, M.J., Wuyts, I.J., Spijkers, W.A.C., and Wallmeyer, P. Gender differences in choice reaction time: Evidence for differential strategies. *Ergonomics* 42(2): 327–335, 1999.

Adams, J.A., Humes, J.M., and Stenson, H.H. Monitoring of complex visual displays: III. Effects of repeated session on human vigilance. *Human Factors* 4(3): 149–158, 1962.

Akamatsu, M., MacKenzie, I.S., and Hasbroucq, T. A comparison of tactile, auditory, and visual feedback in a pointing task using a mouse-type device. *Ergonomics* 38(4), 816–827, 1995.

Alfred, P.E., and Rice, V.J. Age differences in simple and procedural reaction time among healthy military personnel. Proceedings of the Human Factors and Ergonomics Society 56th Annual Meeting, 1809–1813, 2012.

Arrabito, G.R., Ho, G., Li, Y., Giang, W., Burns, C.M., Hou, M., and Pace, P. Proceedings of the Human Factors and Ergonomics Society 57th Annual Meeting, 1164–1168, 2013.

Aykin, N., Czaja, S.J., and Drury, C.G. A simultaneous regression model for double stimulation tasks. *Human Factors* 28(6), 633–643, 1986.

Baker, C.A., Morris, D.F., and Steedman, W.C. Target recognition on complex displays. *Human Factors* 2(2): 51–61, 1960.

Begault, D.R., and Pittman, M.T. Three-dimensional audio versus head-down traffic alert and collision avoidance system displays. *International Journal of Aviation Psychology* 6(1): 79–93, 1996.

Beh, H.C., and Hirst, R. Performance on driving-related tasks during music. *Ergonomics* 42(8): 1087–1098, 1999.

Bemis, S.V., Leeds, J.L., and Winer, E.A. Operator performance as a function of type of display: Conventional versus perspective. *Human Factors* 30(2): 163–169, 1988.

Boehm-Davis, D.A., Holt, R.W., Koll, M., Yastrop, G., and Peters, R. Effects of different data base formats on information retrieval. *Human Factors* 31(5): 579–592, 1989.

Boer, L.C. *Psychological Fitness of Leopard I-V Crews after a 200-km Drive (Report Number 1ZF 1987-30)*. Soesterberg, Netherlands: TNO Institute for Perception, 1987.

Boer, L.C., Ruzius, M.H.B., Minpen, A.M., Bles, W., and Janssen, W.H. *Psychological Fitness during a Maneuver (Report Number 1ZF 1984-17)*. Soesterberg, Netherlands: TNO Institute for Perception, 1984.

Briggs, R.W., and Goldberg, J.H. Battlefield recognition of armored vehicles. *Human Factors* 37(3): 596–610, 1995.

Brown, S., Matsangas, P., and Shattuck, N.L. Comparison of a circadian-based and a forward rotating watch schedules on sleep, mood, and psychomotor vigilance performance. Proceedings of the Human Factors and Ergonomics Society Annual Meeting, 1167–1171, 2015.

Buttigieg, M.A., and Sanderson, P.M. Emergent features in visual display design for two types of failure detection tasks. *Human Factors* 33(6): 631–651, 1991.

Carter, R.C., Krause, M., and Harbeson, M.M. Beware the reliability of slope scores for individuals. *Human Factors* 28(6): 673–683, 1986.

Chancey, E.T., Brill, J.C., Sitz, A., Schmuntzsch, U., and Bliss, J.P. Vibrotactile stimuli parameters on detection reaction times. Proceedings of the Human Factors and Ergonomics Society 58th Annual Meeting, 1701–1705, 2014.

Chapanis, A., and Lindenbaum, L.E. A reaction time study of four control-display linkages. *Human Factors* 1(4): 1–7, 1959.

Chiles, W.D. Effects of elevated temperatures on performance of a complex mental task. *Ergonomics* 2(1): 89–107, 1958.

Claypoole, V.L., and Szalma, J.L. Social norms and cognitive performance: A look at the vigilance decrement in the presence of supervisors. Proceedings of the Human Factors and Ergonomics Society 59th Annual Meeting, 1012–1016, 2015.

Cole, B.L., Johnston, A.W., Gibson, A.J., and Jacobs, R.J. Visual detection of commencement of aircraft takeoff runs. *Aviation, Space, and Environmental Medicine* 49(2): 395–405, 1978.

Coury, B.G., Boulette, M.D., and Smith, R.A. Effect of uncertainty and diagnosticity on classification of multidimensional data with integral and separable displays of system status. *Human Factors* 31(5): 551–569, 1989.

Crawford, A. The perception of light signals: The effect of the number of irrelevant lights. *Ergonomics* 5(3): 417–428, 1962.

Curtis, M.T., Maraj, C., Ritman, M., and Jentsch, F. Investigation of the impact of feedback on decision accuracy and reaction time in a perceptual training task. Proceedings of the Human Factors and Ergonomics Society 54th Annual Meeting, 1630–1634, 2010.

Damos, D. The effect of asymmetric transfer and speech technology on dual-task performance. *Human Factors* 27(4): 409–421, 1985.

Dattel, A.R., Vogt, J.E., Fratzola, J.K., Dever, D.P., Stefonetti, M., Sheehan, C.C., Miller, M.C., and Cavanaugh, J.A. The gorilla's role in relevant and irrelevant stimuli in situation awareness and driving hazard detection. Proceedings of the Human Factors and Ergonomics Society 55th Annual Meeting, 924–928, 2011.

DeMaio, J., Parkinson, S.R., and Crosby, J.V. A reaction time analysis of instrument scanning. *Human Factors* 20(4): 467–471, 1978.

Deupree, R.H., and Simon, J.R. Reaction time and movement time as a function of age, stimulus duration, and task difficulty. *Ergonomics* 6(4): 403–412, 1963.

Dewar, R.E., Ellis, J.G., and Mundy, G. Reaction time as an index of traffic sign perception. *Human Factors* 18(4): 381–392, 1976.

Donders, F.C. On the speed of mental processes. In W.G. Koster (Ed.) *Attention and Performance* (pp. 412–431). Amsterdam: North Holland, 1969.

Donmez, B., Cummings, M.L., and Graham, H.D. Auditory decision aiding in supervisory control of multiple unmanned aerial vehicles. *Human Factors* 51(5): 718–729, 2009.

Downing, J.V., and Sanders, M.S. The effects of panel arrangement and focus of attention on performance. *Human Factors* 29(5): 551–562, 1987.

Elvers, G.C., Adapathya, R.S., Klauer, K.M., Kancler, D.E., and Dolan, N.J. Effects of task probability on integral and separable task performance. *Human Factors* 35(4): 629–637, 1993.

Enander, A. Effects of moderate cold on performance of psychomotor and cognitive tasks. *Ergonomics* 30(10): 1431–1445, 1987.

Finomore, V., Popik, D., Castle, C., and Dallman, R. Effects of a network-centric multimodal communication tool on a communication monitoring task. Proceedings of the Human Factors and Ergonomics Society 54th Annual Meeting, 2125–2129, 2010.

Fisk, A.D., and Jones, C.D. Global versus local consistency: Effects of degree of within-category consistency on performance and learning. *Human Factors* 34(6): 693–705, 1992.

Fowler, B., Elcombe, D.D., Kelso, B., and Portlier, G. The thresholds for hypoxia effects on perceptual-motor performance. *Human Factors* 29(1): 61–66, 1987.

Fowler, B., and Kelso, B. The effects of hypoxia on components of the human event-related potential and relationship to reaction time. *Aviation, Space, and Environmental Medicine* 63(6): 510–516, 1992.

Fowler, B., Mitchell, I., Bhatia, M., and Portlier, G. Narcosis has additive rather than interactive effects on discrimination reaction time. *Human Factors* 31(5): 571–578, 1989.

Fowler, B., Paul, M., Porlier, G., Elcombe, D.D., and Taylor, M. A re-evaluation of the minimum altitude at which hypoxic performance decrements can be detected. *Ergonomics* 28(5): 781–791, 1985.

Fowler, B., White, P.L., Wright, G.R., and Ackles, K.N. The effects of hypoxia on serial response time. *Ergonomics* 25(3): 189–201, 1982.

Frowein, H.W. Selective drug effects on information processing. Dissertatie, Katholieke Hogeschool, Tilburg, 1981.

Frowein, H.W., Gaillard, A.W.K., and Varey, C.A. EP components, visual processing stages, and the effect of a barbiturate. *Biological Psychology* 13: 239–249, 1981a.

Frowein, H.W., Reitsma, D., and Aquarius, C. Effects of two counteractivity stresses on the reaction process. In J. Long and A.D. Baddeley (Eds.) *Attention and Performance*. Hillsdale, NJ: Erlbaum, 1981b.

Gaillard, A.W.K., Gruisen, A., and de Jong, R. *The Influence of Loratadine (sch 29851) on Human Performance (Report Number IZF 1986-C-19)*. Soesterberg, Netherlands: TNO Institute for Perception, 1986.

Gaillard, A.W.K., Rozendaal, A.H., and Varey, C.A. *The Effects of Marginal Vitamin Deficiency on Mental Performance (Report Number IZF 1983-29)*. Soesterberg, Netherlands: TNO Institute for Perception, 1983.

Gaillard, A.W.K., Varey, C.A., and Ruzius, M.H.B. *Marginal Vitamin Deficiency and Mental Performance (Report Number 1ZF 1985-22)*. Soesterberg, Netherlands: TNO Institute for Perception, 1985.

Gliner, J.A., Matsen-Twisdale, J.A., and Horvath, S.M. Auditory and visual sustained attention during ozone exposure. *Aviation, Space, and Environmental Medicine* 50(9): 906–910, 1979.

Gundry, A.J. Experiments on the detection of roll motion. *Aviation, Space, and Environmental Medicine* 49(5): 657–664, 1978.

Haas, E.C., and Casali, J.G. Perceived urgency of and response time to multi-tone and frequency modulated warning signals in broadband noise. *Ergonomics* 38(11): 2313–2326, 1995.

Haines, R.F. Detection time to a point source of light appearing on a star field background with and without a glare source present. *Human Factors* 10(5): 523–530, 1968.

Hancock, P.A., and Caird, J.K. Experimental evaluation of a model of mental workload. *Human Factors* 35(3): 413–429, 1993.

Harris, W.C., Hancock, P.A., Arthur, E.J., and Caird, J.K. Performance, workload, and fatigue changes associated with automation. *International Journal of Aviation Psychology* 5(2): 169–185, 1995.

Heimstra, N.W., Fallesen, J.J., Kinsley, S.A., and Warner, N.W. The effects of deprivation of cigarette smoking on psychomotor performance. *Ergonomics* 23(11): 1047–1055, 1980.

Helton, W.S., and Head, J. Earthquakes on the mind: Implications of disasters for human performance. *Human Factors* 54(2): 189–194, 2012.

Helton, W.S., Russell, P.N., and Dorahy, M.J. Dissociative tendencies and dual-task load: Effects on vigilance performance. Proceedings of the Human Factors and Ergonomics Society 59th Annual Meeting, 711–715, 2015.

Hess, A.S., Wismer, A.J., Bohil, C.J., and Neider, M.B. On the hunt: Visual search for camouflaged targets in realistic environments. Proceedings of the Human Factors and Ergonomics Society 57th Annual Meeting, 1124–1128, 2013.

Hilgendorf, L. Information input and response time. *Ergonomics* 9(1): 31–38, 1966.

Ho, G., Scialfi, C.T., Caird, J.K., and Graw, T. Visual search for traffic signs: The effects of clutter, luminance, and aging. *Human Factors* 43(2): 194–207, 2001.
Holahan, C.J., Culler, R.E., and Wilcox, B.L. Effects of visual detection on reaction time in a simulated traffic environment. *Human Factors* 20(4): 409–413, 1978.
Hollands, J.G., and Lamb, M. Viewpoint tethering for remotely operated vehicles: Effects on complex terrain navigation and spatial awareness. *Human Factors* 53(2): 154–167, 2011.
Horne, J.A., and Gibbons, H. Effects of vigilance performance and sleepiness of alcohol given in the early afternoon ('post lunch') vs. early evening. *Ergonomics* 34(1): 67–77, 1991.
Imbeau, D., Wierwille, W.W., Wolf, L.D., and Chun, G.A. Effects of instrument panel luminance and chromaticity on reading performance and preference in simulated driving. *Human Factors* 31(2): 147–160, 1989.
Jeeves, M.A. Changes in performance at a serial-reaction task under conditions of advance and delay of information. *Ergonomics* 4(4): 327–338, 1961.
Jipp, M. Reaction times to consecutive automation failures: A function of working memory and sustained attention. *Human Factors* 58(8): 1248–1261, 2016.
Johnston, D.M. Target recognition on TV as a function of horizontal resolution and shades of gray. *Human Factors* 10(3): 201–210, 1968.
Jubis, R.M. Coding effects on performance in a process control task with uniparameter and multiparameter displays. *Human Factors* 32(3): 287–297, 1990.
Kerstholt, J.H., Passenier, P.O., Houltuin, K., and Schuffel, H. The effect of a priori probability and complexity on decision making in a supervisory task. *Human Factors* 38(1): 65–78, 1996.
Kimchi, R., Gopher, D., Rubin, Y., and Raij, D. Performance under dichoptic versus binocular viewing conditions: Effects of attention and task requirements. *Human Factors* 35(1): 35–56, 1993.
Kline, T.J.B., Ghali, L.M., Kline, D., and Brown, S. Visibility distance of highway signs among young, middle-aged, and older observers: Icons are better than text. *Human Factors* 32(5): 609–619, 1990.
Koelega, H.S., Brinkman, J., Hendriks, L., and Verbaten, M.N. Processing demands, effort, and individual differences in four different vigilance tasks. *Human Factors* 31(1): 45–62, 1989.
Korteling, J.E. Perception-response speed and driving capabilities of brain-damaged and older drivers. *Human Factors* 32(1): 95–108, 1990.
Krantz, J.H., Silverstein, L.D., and Yeh, Y. Visibility of transmissive liquid crystal displays under dynamic lighting conditions. *Human Factors* 34(5): 615–632, 1992.
Krause, M. Repeated measures on a choice reaction time task. Proceedings of the Human Factors Society 26th Annual Meeting, 359–363, 1982.
Lee, M.D., and Fisk, A.D. Disruption and maintenance of skilled visual search as a function of degree of consistency. *Human Factors* 35(2): 205–220, 1993.
Leifflen, D., Poquin, D., Savourey, G., Barraud, P., Raphel, C., and Bittel, J. Cognitive performance during short acclimation to severe hypoxia. *Aviation, Space, and Environmental Medicine* 68(11): 993–997, 1997.
Leonard, J.A., and Carpenter, A. On the correlation between a serial choice reaction task and subsequent achievement at typewriting. *Ergonomics* 7(2): 197–204, 1964.
Liebowitz, H.W., Abernathy, C.N., Buskirk, E.R., Bar-or, O., and Hennessy, R.T. The effect of heat stress on reaction time to centrally and peripherally presented stimuli. *Human Factors* 14(2): 155–160, 1972.

Linden, A.E., Nathoo, A., and Fowler, B. An AFM investigation of the effects of acute hypoxia on mental rotation. *Ergonomics* 39(2): 278–284, 1996.

Loeb, M., and Schmidt, E.A. A comparison of the effects of different kinds of information in maintaining efficiency on an auditory monitoring task. *Ergonomics* 6(1): 75–82, 1963.

Logsdon, R., Hochhaus, L., Williams, H.L., Rundell, O.H., and Maxwell, D. Secobarbital and perceptual processing. *Acta Psychologica* 55: 179–193, 1984.

MacDonald, W.A., and Cole, B.L. Evaluating the role of colour in a flight information cockpit display. *Ergonomics* 31(1): 13–37, 1988.

Macintosh, J.H., Thomas, D.J., Olive, J.E., Chesner, I.M., and Knight, R.J.E. The effect of altitude on tests of reaction time and alertness. *Aviation, Space, and Environmental Medicine* 59(3): 246–248, 1988.

Mackie, R.R., and Wylie, C.D. Countering loss of vigilance in sonar watch standing using signal injection and performance feedback. *Ergonomics* 37(7): 1157–1184, 1994.

Matthews, G., and Ryan, H. The expression of the 'pre-menstrual' syndrome in measures of mood and sustained attention. *Ergonomics* 37(8): 1407–1417, 1994.

McCarthy, D., Corban, R., Legg, S., and Faris, J. Effects of mild hypoxia on perceptual-motor performance: A signal-detection approach. *Ergonomics* 38(10): 1979–1992, 1995.

McKnight, A.J., and Shinar, D. Brake reaction time to center high-mounted stop lamps on vans and trucks. *Human Factors* 34(2): 205–213, 1992.

Mertens, H.W., and Collins, W.E. The effects of age, sleep deprivation, and altitude on complex performance. *Human Factors* 28(5): 541–551, 1986.

Miles, C., Auburn, T.C., and Jones, D.M. Effects of loud noise and signal probability on visual vigilance. *Ergonomics* 27(8): 855–862, 1984.

Moacdieh, N.M., Prinet, J.C., and Sarter, N.B. Effects of modern primary flight display clutter: Evidence from performance and eye tracking data. Proceedings of the Human Factors and Ergonomics Society 57th Annual Meeting, 11–15, 2013.

Moacdieh N., and Sarter, N. Data density and poor organization: Analyzing the performance and attentional effects of two aspects of display clutter. Proceedings of the Human Factors and Ergonomics Society 59th Annual Meeting, 1336–1340, 2015.

Moraal, J. *Age and Information Processing: An Application of Sternberg's Additive Factor Method (Report Number 1ZF 1982-18).* Soesterberg, Netherlands: TNO Institute for Perception, 1982.

Morgerstern, F.S., and Haskell, S.H. Disruptive reaction times in single and multiple response units. *Ergonomics* 14(2): 219–230, 1971.

Murray, S.A., and Caldwell, B.S. Human performance and control of multiple systems. *Human Factors* 38(2): 323–329, 1996.

Nataupsky, M., and Crittenden, L. Stereo 3-D and non-stereo presentations of a computer-generated pictorial primary flight display with pathway augmentation. Proceedings of the 9th AIAA/IEEE Digital Avionics Systems Conference, 1988.

Norman, J., and Ehrlich, S. Visual accommodation and virtual image displays: Target detection and recognition. *Human Factors* 28(2): 135–151, 1986.

Park, K.S., and Lee, S.W. A computer-aided aptitude test for predicting flight performance of trainees. *Human Factors* 34(2): 189–204, 1992.

Pascale, M., Sanderson, P., Liu, D., Mohamed, I., Stigter, N., and Loeb, R. Peripheral detection for abrupt onset stimuli presented via head-worn display. Proceedings of the Human Factors and Ergonomics Society 59th Annual Meeting, 1326–1330, 2015.

Payne, D.G., and Lang, V.A. Visual monitoring with spatially versus temporally distributed displays. *Human Factors* 33(4): 443–458, 1991.

Perriment, A.D. The effect of signal characteristics on reaction time using bisensory stimulation. *Ergonomics* 12(1): 71–78, 1969.

Perrott, D.R., Sadralodabai, T., Saberi, K., and Strybel, T.Z. Aurally aided visual search in the central visual field: Effects of visual load and visual enhancement of the target. *Human Factors* 33(4): 389–400, 1991.

Phillips, J.B., Simmons, R.G., Florian, J.P., Horning, D.S., Lojewski, R.A., and Chandler, J.F. Moderate intermittent hypoxia: Effect on two-choice reaction time followed by a significant delay in recovery. Proceedings of the Human Factors and Ergonomics Society 53rd Annual Meeting, 1564–1568, 2009.

Pigeau, R.A., Angus, R.G., O'Neill, P., and Mack, I. Vigilance latencies to aircraft detection among NORAD surveillance operators. *Human Factors* 37(3): 622–634, 1995.

Ralston, J.V., Pisoni, D.B., Lively, S.E., Greene, B.G., and Mullennix, J.W. Comprehension of synthetic speech produced by rule: Word monitoring and sentence-by-sentence listening times. *Human Factors* 33(4): 471–491, 1991.

Remington, R., and Williams, D. On the selection and evaluation of visual display symbology: Factors influencing search and identification times. *Human Factors* 28(4): 407–420, 1986.

Rizzardo, C.A., and Colle, H.A. Dual-coded advisory turn indicators for GPS navigational guidance of surface vehicles: Effects of map orientation. *Human Factors* 55(5): 935–945, 2013.

Robinson, E., Simpson, B., Finomore, V., Cowgill, J., Shalin, V.L., Hampton, A., Moore, T., Rapoch, T., and Gilkey, R. Aurally aided visual threat acquisition in a virtual urban environment. Proceedings of the Human Factors and Ergonomics Society 56th Annual Meeting, 1471–1475, 2012.

Sanders, A.F., Wijnen, J.I.C., and van Arkel, A.E. An additive factor analysis of the effects of sleep-loss on reaction processes. *Acta Psychologica* 51: 41–59, 1982.

Shattuck, N.L., Shattuck, L.G., Smith, K., and Matsangas, P. Changes in reaction times and executive decision-making following exposure to waterborne motion. Proceedings of the Human Factors and Ergonomics Society 57th Annual Meeting, 1987–1991, 2013.

Shvartz, E., Meroz, A., Mechtinger, A., and Birnfeld, H. Simple reaction time during exercise, heat exposure, and heat acclimation. *Aviation, Space, and Environmental Medicine* 47(11): 1168–1170, 1976.

Simon, J.R. Choice reaction time as a function of auditory S-R correspondence, age, and sex. *Ergonomics* 10(6): 659–664, 1967.

Simon, J.R., and Overmeyer, S.P. The effect of redundant cues on retrieval time. *Human Factors* 26(3): 315–321, 1984.

Simon, J.R., Peterson, K.D., and Wang, J.H. Same-different reaction time to stimuli presented simultaneously to separate cerebral hemispheres. *Ergonomics* 31(12): 1837–1846, 1988.

Simon, J.R., and Wolf, J.D. Choice reaction time as a function of angular stimulus-response correspondence and age. *Ergonomics* 6(1): 99–106, 1963.

Steelman, K.S., and McCarley, J.S. Interaction among target salience, eccentricity, target expectancy and workload in an alert detection task. Proceedings of the Human Factors and Ergonomics Society 55th Annual Meeting, 1407–1411, 2011.

Steiner, B.A., and Camacho, M.J. Situation awareness: Icons vs. alphanumerics. Proceedings of the Human Factors Society 33rd Annual Meeting, 28–32, 1989.

Stelzer, E.M., and Klein, K.A. Effectiveness of a spatial algorithm for air traffic controller use in airport surface conformance monitoring. Proceedings of the Human Factors Society 55th Annual Meeting, 6–9, 2011.

Stewart, J.D., and Clark, B. Choice reaction time to visual motion during prolonged rotary motion in airline pilots. *Aviation, Space, and Environmental Medicine* 46(6): 767–771, 1975.

Steyvers, F.J.J.M. The influence of sleep deprivation and knowledge of results on perceptual encoding. *Acta Psychologica* 66: 173–178, 1987.

Swanson, L., Jones, E., Riordan, B., Bruni, S., Schurr, N., Sullivan, S., and Lansey, J. Exploring human error in an RPA target detection task. Proceedings of the Human Factors and Ergonomics Society 56th Annual Meeting, 328–332, 2012.

Swink, J.R. Intersensory comparisons of reaction time using an electro-pulse tactile stimulus. *Human Factors* 8(2): 143–146, 1966.

Taylor, R.M., and Selcon, S.J. Cognitive quality and situational awareness with advanced aircraft attitude displays. Proceedings of the Human Factors Society 34th Annual Meeting, 26–30, 1990.

Tear, C.L., Fox, M., Tsai, M., Liu, D., and Sanderson, P.M. Detecting numerical and waveform changes on a head-mounted display vs. monitor. Proceedings of the Human Factors and Ergonomics Society 57th Annual Meeting, 1134–1138, 2013.

Teichner, W.H. Probability of detection and speed of response in simple monitoring. *Human Factors* 4(4): 181–186, 1962.

Teo, G.W.L., and Szalma, J.L. The effect of spatial and temporal task characteristics on performance, workload, and stress. Proceedings of the Human Factors and Ergonomics Society 54th Annual Meeting, 1699–1703, 2010.

Thackray, R.I., Bailey, J.P., and Touchstone, R.M. The effect of increased monitoring load on vigilance performance using a simulated radar display. *Ergonomics* 22(5): 529–539, 1979.

Thackray, R.I., and Touchstone, R.M. Effects of monitoring under high and low task load on detection of flashing and coloured radar targets. *Ergonomics* 34(8): 1065–1081, 1991.

Thackray, R.I., Touchstone, R.M., and Bailey, J.P. Comparison of the vigilance performance of men and women using a simulated radar task. *Aviation, Space, and Environmental Medicine* 49(10): 1215–1218, 1978.

Tullis, T.S. Human performance evaluation of graphic and textual CRT displays of diagnostic data. Proceedings of the Human Factors Society 24th Annual Meeting, 310–316, 1980.

Tzelgov, J., Henik, A., Dinstein, I., and Rabany, J. Performance consequence of two types of stereo picture compression. *Human Factors* 32(2): 173–182, 1990.

Van Assen, A., and Eijkman, E.G. Reaction time and performance in learning a two dimensional compensatory tracking task. *Ergonomics* 13(9): 707–717, 1971.

Vidulich, M.A. The Bedford Scale: Does it measure spare capacity? Proceedings of the 6th International Symposium on Aviation Psychology, 1136–1141, 1991.

Warner, H.D., and Heimstra, N.W. Effects of noise intensity on visual target-detection performance. *Human Factors* 14(2): 181–185, 1972.

Warrick, M.J., Kibler, A.W., and Topmiller, D.A. Response time to unexpected stimuli. *Human Factors* 7(1): 81–86, 1965.

Welford, A.T. What is the basis of choice reaction-time? *Ergonomics* 14(6): 679–693, 1971.

Wickens, C.D., and Ward, J. Cockpit displays of traffic and weather information: Effects of 3D perspective versus 2D coplanar rendering and database integration. *International Journal of Aerospace Psychology* 27 (1–2): 44–56, 2017.

Wierwille, W.W., Casali, J.G., and Repa, B.S. Driver steering reaction time to abrupt-onset crosswinds, as measured in a moving-base driving simulator. *Human Factors* 25(1): 103–116, 1983.

Wierwille, W.W., Rahimi, M., and Casali, J.G. Evaluation of 16 measures of mental workload using a simulated flight task emphasizing mediational activity. *Human Factors* 27(5): 489–502, 1985.

Wolcott, J. H., Hanson, C.A., Foster, W.D., and Kay, T. Correlation of choice reaction time performance with biorhythmic criticality and cycle phase. *Aviation, Space, and Environmental Medicine* 50(1): 34–39, 1979.

Yeh, Y., and Silverstein, L.D. Spatial judgments with monoscopic and stereoscopic presentation of perspective displays. *Human Factors* 34(5): 583–600, 1992.

Zeiner, A.R., and Brecher, G.A. Reaction time with and without backscatter from intense pulsed light. *Aviation, Space, and Environmental Medicine* 46(2): 125–127, 1975.

2.2.7 Reading Speed

General description – Reading speed is the number of words read divided by the reading time interval. Reading speed is typically measured in words per minute.

Strengths and limitations – In an early study, Seminar (1960) measured reading speed for tactually presented letters. He reported an average speed of 5.5 seconds for two letters and 25.5 seconds for seven letters. The participants were three men and three women.

Cushman (1986) reported that reading speeds tend to be slower for negative than for positive images. Since there may be a speed-accuracy trade-off, Cushman (1986) also calculated overall reading performance (reading speed × percentage of reading comprehension questions answered correctly). In the same year, Gould and Grischkowsky (1986) reported that reading speed decreases as visual angles increase over 24.3 degrees. In a follow-on study, Gould et al. (1987) reported in a series of 10 experiments that reading speed is slower from cathode ray tube (CRT) displays than from paper. In a similar study, Jorna and Snyder (1991) reported equivalent reading speeds for hard copy and soft copy displays if the image qualities are similar.

Gould et al. (1987) concluded on the basis of six studies that reading speed was equivalent on paper and CRT if the CRT displays contained "character fonts that resemble those on paper (rather than dot matrix fonts, for example), that have a polarity of dark characters on a light background, that are anti-aliased (i.e., contain grey level), and that are shown on displays with relatively high resolution (e.g., 1,000 × 800)" (p. 497).

A year later, Chen et al. (1988) used the average reading rate, in words per minute (wpm) to evaluate the effects of window size (20 versus 40 characters) and jump length (i.e., number of characters that a message is advanced horizontally) of a visual display. These authors reported that reading rate was significantly less for one-jump (90–91 wpm) than for five- (128 wpm) and nine-jump (139–144 wpm) conditions. Reading rate was not significantly affected by window size, however.

Campbell et al. (1981) reported increased reading speed for justified rather than unjustified text. Moseley and Griffin (1986) reported increased reading time and reading error in vibration of the display, the participant, or both.

Lachman (1989) used the inverse of reading time, i.e., reading rate, to evaluate the effect of presenting definitions concurrently with text on a CRT display. There was a significantly higher reading rate for the first 14 screens read than for the second 14 screens.

Data requirements – The number of words and the duration of the reading interval must be recorded.

Thresholds – Cushman (1986) reported the following average words/minute: paper = 218; positive image microfiche = 210; negative image microfiche = 199; positive image, negative contrast Video Display Terminal (VDT) = 216; and negative image, positive contrast VDT = 209.

Sources

Campbell, A.J., Marchetti, F.M., and Mewhort, D.J.K. Reading speed and text production A note on right-justification techniques. *Ergonomics* 24(8): 633–640, 1981.

Chen, H., Chan, K., and Tsoi, K. Reading self-paced moving text on a computer display. *Human Factors* 30(3): 285–291, 1988.

Cushman, W.H. Reading from microfiche, a VDT, and the printed page: Subjective fatigue and performance. *Human Factors* 28(1): 63–73, 1986.

Gould, J.D., Alfaro, L., Finn, R., Haupt, B., and Minuto, A. Reading from CRT displays can be as fast as reading from paper. *Human Factors* 29(5): 497–517, 1987.

Gould, J.D., and Grischkowsky, N. Does visual angle of a line of characters affect reading speed? *Human Factors* 28(2): 165–173, 1986.

Jorna, G.C., and Snyder, H.L. Image quality determines differences in reading performance and perceived image quality with CRT and hard-copy displays. *Human Factors* 33(4): 459–469, 1991.

Lachman, R. Comprehension aids for on-line reading of expository text. *Human Factors* 31(1): 1–15, 1989.

Moseley, M.J., and Griffin, M.J. Effects of display vibration and whole-body vibration on visual performance. *Ergonomics* 29(8): 977–983, 1986.

Seminar, J.L. Accuracy and speed of tactual reading: An exploratory study. *Ergonomics* 3(1): 62–67, 1960.

2.2.8 Search Time

General description – Search time is the length of time for a user to retrieve the desired information from a database. Lee and MacGregor (1985) provided the following definition.

$$\mathrm{st} = r(at + k + c)$$

where

st = search time
r = total number of index pages accessed in retrieving a given item
a = number of alternatives per page
t = time required to read one alternative
k = key-press time
c = computer response time (pp. 158–159)

Matthews (1986) defined search time as the length of time for a participant to locate and indicate the position of a target.

Strengths and limitations – Search time has been used to evaluate displays, clutter, and time on task.

Displays. Fisher et al. (1989) used search time to evaluate the effect of highlighting on visual displays. Search time was significantly longer when the probability that the target was highlighted was low (0.25) rather than high (0.75). This is similar to an earlier study by Monk (1976) in which target uncertainty increased search time by 9.5%.

In another display study, Harpster et al. (1989) reported significantly longer search times using a low resolution/addressability ratio (RAR) than using a high RAR or hard copy. Vartabedian (1971) reported a significant increase in search time for lowercase words than for uppercase words. Matthews et al. (1989) reported significantly longer search times for green on black displays (7.71 s) than red on black displays (7.14 s). Hollands et al. (2002) reported significantly longer search times for diamond shapes rather than square shapes on a CRT. There was no difference, however, on a Liquid Crystal Display (LCD). There were, however, longer search times on the LCD for red and blue symbols than for white symbols. There were no color effects on the CRT.

Erickson (1964) reported few significant correlations between search time and peripheral visual acuity. There was a significant interaction of target shape and number of objects in a display: rings were found in significantly shorter times than blobs when there were fewer objects in a display (16 versus 48). Siva et al. (2014) calculated search efficiency slopes using search time and set size to evaluate standardized symbology in MIL STD 2525. They reported significant effects of feature and distinctiveness.

Clutter. Brown and Monk (1975) reported longer visual search times as the number of nontargets increased and as the randomness of the location of the nontargets increased. Matthews (1986) reported that search time in the current trial was significantly increased if the visual load of the previous trial was high. Nagy and Sanchez (1992) used mean log search time to investigate the effects of luminance and chromaticity differences between targets and distractors. "Results showed that mean search time increased linearly with the number of distractors if the luminance difference between target and distractors was small but was roughly constant if the luminance difference was large" (p. 601).

Bednall (1992) reported shorter search times for spaced rather than nonspaced targets, targets with alternating lines rather than non-alternating lines, and insertions of blank lines. There was no effect of all capital versus mixed case letters. Lee and MacGregor (1985) reported that their search time model was evaluated using a videotex information retrieval system. It was useful in evaluating menu design decisions. McIntire et al. (2010) compared the time to locate a single visual target among 15 visual nontargets with and without 3-D auditory cues and in static or dynamic environments. Auditory cues reduced search times by 22% in static environments and 25% in dynamic environments.

Time on Task. Lovasik et al. (1989) reported significantly longer search times in the first half hour than in the remaining three and a half hours of a visual search task.

Carter et al. (1986) reported that RT was more reliable than slope for this task.

Data requirement – User search time can be applied to any computerized database in which the parameters a, c, k, r, and t can be measured.

Thresholds – Not stated.

Sources

Bednall, E.S. The effect of screen format on visual list search. *Ergonomics* 35(4): 369–383, 1992.

Brown, B., and Monk, T.H. The effect of local target surround and whole background constraint on visual search times. *Human Factors* 17(1): 81–88, 1975.

Carter, R.C., Krause, M., and Harbeson, M.M. Beware the reliability of slope scores for individuals. *Human Factors* 28(6): 673–683, 1986.

Erickson, R.A. Relation between visual search time and peripheral visual acuity. *Human Factors* 6(2): 165–178, 1964.

Fisher, D.L., Coury, B.G., Tengs, T.O., and Duffy, S.A. Minimizing the time to search visual displays: The role of highlighting. *Human Factors* 31(2): 167–182, 1989.

Harpster, J.K., Freivalds, A., Shulman, G.L., and Leibowitz, H.W. Visual performance on CRT screens and hard-copy displays. *Human Factors* 31(3): 247–257, 1989.

Hollands, J.G., Parker, H.A., McFadden, S., and Boothby, R. LCD versus CRT displays: A comparison of visual search performance for colored symbols. *Human Factors* 44(2): 210–221, 2002.

Lee, E., and MacGregor, J. Minimizing user search time in menu retrieval systems. *Human Factors* 27(2): 157–162, 1985.
Lovasik, J.V., Matthews, M.L., and Kergoat, H. Neural, optical, and search performance in prolonged viewing of chromatic displays. *Human Factors* 31(3): 273–289, 1989.
Matthews, M.L. The influence of visual workload history on visual performance. *Human Factors* 28(6): 623–632, 1986.
Matthews, M.L., Lovasik, J.V., and Mertins, K. Visual performance and subjective discomfort in prolonged viewing of chromatic displays. *Human Factors* 31(3): 259–271, 1989.
McIntire, J.P., Havig, P.R., Watamaniuk, S.N.J., and Gilkey, R.H. Visual search performance with 3-D auditory cues: Effects of motion, target location, and practice. *Human Factors* 52(1): 41–53, 2010.
Monk, T.H. Target uncertainty in applied visual search. *Human Factors* 18(6): 607–612, 1976.
Nagy, A.L., and Sanchez, R.R. Chromaticity and luminance as coding dimensions in visual search. *Human Factors* 34(5): 601–614, 1992.
Siva, N., Chaparro, A., and Palmer, E. Evaluation of MILSTD 2525 glyph features in a visual search paradigm. Proceedings of the Human Factors and Ergonomics Society 58th Annual Meeting, 1189–1193, 2014.
Vartabedian, A.G. The effects of letter size, case, and generation method on CRT display search time. *Human Factors* 13(4): 363–368, 1971.

2.2.9 Task Load

General description – Task load is the time required to perform a task divided by the time available to perform the task. Values above 1 indicate excessive task load.

Strengths and limitations – Task load is sensitive to workload in in-flight environments. For example, Geiselhart et al. (1976) used task load to identify differences in workload among four types of refueling missions. Geiselhart et al. (1977) used task load to estimate the workload of KC-135 crews. Using this method, these researchers were able to quantify differences in task load between different types of missions and crew positions. Gunning and Manning (1980) calculated the percentage of time spent on each task for three crewmembers during an aerial refueling. They reported the following inactivity percentages by crew position: pilot, 5 percent; copilot, 45 percent; navigator, 65 percent. Task load was high during takeoff, air refueling, and landing.

Stone et al. (1984), however, identified three problems with task load: "(1) It does not consider cognitive or mental activities. (2) It does not take into account variations associated with ability and experience or dynamic, adaptive behavior. (3) It cannot deal with simultaneous or continuous-tracking tasks" (p. 14).

Data requirements – Use of the task load method requires: (1) clear visual and auditory records of participants and (2) objective measurement criteria for identifying the starts and ends of tasks.

Thresholds – Not stated.

Sources

Geiselhart, R., Koeteeuw, R.I., and Schiffler, R.J. A Study of Task Loading Using a Four-Man Crew on a KC-135 Aircraft (Giant Boom) (ASD-TR-76-33). Wright-Patterson Air Force Base, OH: Aeronautical Systems Division, April 1977.

Geiselhart, R., Schiffler, R.J., and Ivey, L.J. *A Study of Task Loading Using a Three Man Crew on a KC-135 Aircraft (ASD-TR-76-19).* Wright-Patterson Air Force Base, OH: Aeronautical Systems Division, October 1976.

Gunning, D., and Manning, M. The measurement of aircrew task loading during operational flights. Proceedings of the Human Factors Society 24th Annual Meeting, 249–252, 1980.

Stone, G., Gulick, R.K., and Gabriel, R.F. *Use of Task/Timeline Analysis to Assess Crew Workload (Douglas Paper 7592).* Longbeach: Douglas Aircraft Company, 1984.

2.2.10 Time to Complete

General description – Time to complete is the duration from the participant's first input to the last response (Casali et al., 1990).

Strengths and limitations – Time to complete has been extensively to evaluate displays, controls, automation, operating conditions, population characteristics, task type, and type of training.

Displays. For unmanned systems, Massimino and Sheridan (1994) reported no difference in task competition times between direct and video viewing during teleoperation. Chen et al. (2010) reported no significant difference in course completion time for teleoperation of a robot between 2D and 3D displays. Hollands and Lamb (2011) in a similar study reported navigation time to be significantly longer in exocentric than in egocentric displays. In a medical application, Sublette et al. (2010) also reported no significant difference between 2D and 3D displays for minimally invasive surgery.

Matheson et al. (2013) reported significant improvements in time to complete a course with a tele-operated rover using predictive displays.

In another unusual study, Luz et al. (2010) and Manzey et al. (2011) used the time to complete a simulated Mastoidectomy to evaluate the effectiveness of image-guided navigation during surgery. They reported that surgical students required significantly less time to complete the procedure manually than with the imagery system. In an aviation study, Milner et al. (2017) compared time to respond to questions related to instrument approaches using either paper or electronic charts. Twenty-seven pilots were significantly faster using the electronic charts.

Controls. Billings and Durlach (2009) reported faster mission completion times using a game controller rather than a mouse. The vehicle being controlled was a micro Unmanned Aerial Vehicle (UAV). Song et al. (2011) reported significantly faster text entry completion time on the second than on the first trial. Further text entry using two hands was faster than text entry using either the right or the left hand alone.

Automation. Adelman et al. (1993) used time to complete an aircraft identification task to evaluate expert system interfaces and capabilities. They reported that operators took longer to examine aircraft with the screening rather than with the override interface. In an unmanned system study, PrinetI et al. (2012) reported significantly faster completion of a re-planning task for UAV missions with full automation than in manual mode. In a fault identification task, Manzey et al. (2009) reported shorter task completion times when an automated decision aid was available than when it was not.

Operating Conditions. In an aviation maintenance experiment, Warren et al. (2013) reported that Aviation Maintenance Technicians completed tasks more quickly when under time pressure. In an editing task, Brand and Judd (1993) reported significant differences in editing time as a function of the angle of hard copy (316.75 s for 30-degree, 325.03 s for 0 degree; and 371.92 s for 90 degree). In a decision-making task, Shattuck et al. (2009) reported significantly longer times for ambiguous and missing information scenarios than for conflicting information scenarios.

Frankish and Noyes (1990) used rate of data entry to evaluate data feedback techniques. They reported significantly higher rates for visual presentation and for visual feedback than for spoken feedback. In a mundane task, Weber et al. (2013) reported significantly faster times to complete an email classification task (spam versus not spam) with low versus high variability in system response times.

Population Characteristics. Casali et al. (1990) reported significant effects of speech recognition system accuracy and available vocabulary but not a significant age effect. There were also several interactions. In an unusual study, Troutwine and O'Neal (1981) reported time was judged shorter on an interesting rather than a boring task but only for participants with volition. For participants without volition there was no significant difference in time estimation between the two types of tasks.

Four studies were completed looking at individual differences. In the first, Hartley et al. (1987) used completion times to assess the effect of menstrual cycle. Times were slower on verbal reasoning involving complex sentences during ovulation than during menstruation and premenstruation. In the second study, Richter and Salvendy (1995) reported that introverted users of a computer program performed faster with interfaces that were perceived as introverted than with interfaces that were perceived as extroverted. Best et al. (1996) reported that persons with very near dark vergence positions before a visual inspection task performed significantly faster than persons with far dark convergence positions. Their participants were 38 university students with a mean age of 20.6 years. Finally, Lee et al. (2010) reported that older participants (>55 years of age) took longer to complete a map search task than younger participants (<55 years of age).

Task Type. In early work, Burger et al. (1970) used experts to estimate the time it would take to perform tasks. The estimates were then tested against the real times. Although the correlation between the estimates and the real times was high (+0.98), estimates of the minimum performance times were higher than the actual times and varied widely between judges. The tasks were throw toggle switch, turn rotary switch to a specified value, push toggle, observe and record data, and adjust dial. In a very complex task, clearing an alarm in a nuclear power plant simulation, Thornburg et al. (2012) reported significant task completion times as a function of alarm onset time.

Type of Training. Abidi et al. (2012) used time to complete car fender assembly to evaluate the effectiveness of three types of training systems (traditional engineering, Computer Aided Design Environment, and immersive virtual reality). The time to complete was significantly longer for the traditional engineering training than for either of the other two methods.

Time to complete provides a measure of task difficulty but may be traded off for accuracy.

Data requirements – The start and end of the task must be well defined.

Thresholds – Minimum time is 30 msec.

Sources

Abidi, M.H., Ahmad, A., El-Tamini, A.M., and Al-Ahmari, A.M. Development and evaluation of a virtual assembly trainer. Proceedings of the Human Factors and Ergonomics Society 56th Annual Meeting, 2560–2564, 2012.

Adelman, L., Cohen, M.S., Bresnick, T.A., Chinnis, J.O., and Laskey, K.B. Real-time expert system interfaces, cognitive processes, and task performance: An empirical assessment. *Human Factors* 35(2): 243–261, 1993.

Best, P.S., Littleton, M.H., Gramopadhye, A.K., and Tyrrell, R.A. Relations between individual differences in oculomotor resting states and visual inspection performance. *Human Factors* 39(1): 35–40, 1996.

Billings, D.R., and Durlach, P.J. Mission completion time is sensitive to teleoperation performance during simulated reconnaissance missions with a micro-unmanned aerial vehicle. Proceedings of the Human Factors and Ergonomics Society 53rd Annual Meeting, 1408–1412, 2009.

Brand, J.L., and Judd, K. Angle of hard copy and text-editing performance. *Human Factors* 35(1): 57–70, 1993.

Burger, W.J., Knowles, W.B., and Wulfeck, J.W. Validity of expert judgments of performance time. *Human Factors* 12(5): 503–510, 1970.

Casali, S.P., Williges, B.H., and Dryden, R.D. Effects of recognition accuracy and vocabulary size of a speech recognition system on task performance and user acceptance. *Human Factors* 32(2): 183–196, 1990.

Chen, J.Y.C., Oden, R.N.V., Kenny, C., and Merritt, J.O. Stereoscopic displays for robot teleoperation and simulated driving. Proceedings of the Human Factors and Ergonomics Society 54th Annual Meeting, 1488–1492, 2010.

Frankish, C., and Noyes, J. Sources of human error in data entry tasks using speech input. *Human Factors* 32(6): 697–716, 1990.

Hartley, L.R., Lyons, D., and Dunne, M. Memory and menstrual cycle. *Ergonomics* 30(1): 111–120, 1987.
Hollands, J.G., and Lamb, M. Viewpoint tethering for remotely operated vehicles: Effects on complex terrain navigation and spatial awareness. *Human Factors* 53(2): 154–167, 2011.
Lee, D., Jeong, C., and Chung, M.K. Effects of user age and zoomable user interfaces on information searching tasks in a map-type space. Proceedings of the Human Factors and Ergonomics Society 54th Annual Meeting, 571–575, 2010.
Luz, M., Mueller, S., Strauss, G., Dietz, A., Meixenberger, J., and Manzey, D. Automation in surgery: The impact of navigation-control assistance on the performance, workload and situation awareness of surgeons. Proceedings of the Human Factors and Ergonomics Society 54th Annual Meeting, 889–893, 2010.
Manzey, D., Luz, M., Mueller, S., Dietz, A., Meixenberger, J., and Strauss, G. Automation in surgery: The impact of navigation-control assistance on performance, workload, situation awareness, and acquisition of surgical skills. *Human Factors* 53(6): 584–599, 2011.
Manzey, D., Reichenbach, J., and Onnasch, L. Human performance consequences of automated decisions aids in states of fatigue. Proceedings of the Human Factors and Ergonomics Society 53rd Annual Meeting, 329–333, 2009.
Massimino, J.J., and Sheridan, T.B. Teleoperator performance with varying force and visual feedback. *Human Factors* 36(1): 145–157, 1994.
Matheson, A., Donmez, B., Rehmatullah, F., Jasiobedzki, P., Ng, H., Panwar, V., and Li, M. The effects of predictive displays on performance in driving tasks with multi-second latency: Aiding tele-operation of lunar rovers. Proceedings of the Human Factors and Ergonomics Society 57th Annual Meeting, 21–25, 2013.
Milner, M., Bush, D., Marte, D., Rice, S., Winter, S., Adkins, E., Roccasecca, A., and Tamilselvan, G. The effect of chart type on pilots' response time. Proceedings of the Human Factors and Ergonomics Society Annual Meeting, 1365–1368, 2017.
Prinet, J.C., Terhune, A., and Sarter, N.B. Supporting dynamic re-planning in multiple UAV control: A comparison of 3 levels of automation. Proceedings of the Human Factors and Ergonomics Society 56th Annual Meeting, 423–427, 2012.
Richter, L.A., and Salvendy, G. Effects of personality and task strength on performance in computerized tasks. *Ergonomics* 38(2): 281–291, 1995.
Shattuck, L.G., Lewis Miller, N., and Kemmerer, K.E. Tactical decision making under conditions of uncertainty: An empirical study. Proceedings of the Human Factors and Ergonomics Society 53rd Annual Meeting, 242–246, 2009.
Song, J., Ryu, T., Bahn, S., and Yun, M.H. Performance analysis of text entry with preferred one hand using smartphone touch keyboard. Proceedings of the Human Factors and Ergonomics Society 55th Annual Meeting, 1289–1292, 2011.
Sublette, M., Carswell, C.M., Han, Q., Grant, R., Lio, C.H., Lee, G., Field, M., Staley, D., Seales, W.B., and Clarke, D. Dual-view displays for minimally invasive surgery: Does the addition of a 3-D global view decrease mental workload? Proceedings of the Human Factors and Ergonomics Society 54th Annual Meeting, 1581–1585, 2010.
Thornburg, K.M., Peterse, H.P.M., and Liu, A.M. Operator performance in long duration control operations switching from low to high task load. Proceedings of the Human Factors and Ergonomics Society 56th Annual Meeting, 2002–2005, 2012.

Troutwine, R., and O'Neal, E.C. Volition, performance of a boring task and time estimation. *Perceptual and Motor Skills* 52: 865–866, 1981.
Warren, W.R., Blickensderfer, B., Cruit, J., and Boquet, A. Shift turnover strategy and time in aviation maintenance. Proceedings of the Human Factors and Ergonomics Society 57th Annual Meeting, 46–50, 2013.
Weber, F., Haering, C., and Thomaschke, R. Improving the human-computer dialog with increased temporal predictability. *Human Factors* 55(5): 881–892, 2013.

2.3 Task Batteries

The third category is task battery. Task batteries are collections of two or more tasks performed in series or in parallel to measure range of abilities or effects. These batteries assume that human abilities vary across types of tasks or are differentially affected by independent variables. Examples include AGARD's Standardized Tests for Research with Environmental Stressors (STRES) Battery (Section 2.3.1), the Armed Forces Qualification Test (Section 2.3.2), Deutsch and Malmborg (1982) Measurement Instrument Matrix (Section 2.3.3), Performance Evaluation Tests for Environmental Research (PETER) (Section 2.3.4), the Work and Fatigue Test Battery (Section 2.3.5), and the Unified Tri-services Cognitive Performance Assessment Battery (UTCPAB) (Section 2.3.6).

2.3.1 AGARD's Standardized Tests for Research with Environmental Stressors (STRES) Battery

General description – The Standardized Tests for Research with Environmental Stressors (STRES) Battery is made up of seven tests:

1. Reaction time,
2. Mathematical processing,
3. Memory search,
4. Spatial processing,
5. Unstable tracking,
6. Grammatical reasoning, and
7. Dual task performance of Tests 3 and 5.

Strengths and limitations – Tests were selected for the STRES Battery based on the following criteria: "(1) preliminary evidence of reliability, validity, and sensitivity, (2) documented history of application to assessment of a range of stressor effects, (3) short duration (maximum of three minutes per trial block), (4) language-independence, (5) sound basis in [Human Performance

Theory] HPT, [and] (6) ability to be implemented on simple and easily-available computer systems" (AGARD, 1989, p. 7).

Data requirements – Each STRES Battery test has been programmed for computer administration. The order of presentation is fixed, as presented above. Standardized instructions are used as well as a standardized data file format. Test stimuli must be presented in white on a dark background. The joystick must have 30-degree lateral travel from the vertical position, friction not greater than 50 g, linear relationship between angular rotation and lateral movement, and 8-bit resolution.

Thresholds – Not stated.

Source

AGARD. *Human Performance Assessment Methods (AGARD-AG-308).* Neuilly-sur-Seine, France: AGARD, June 1989.

2.3.2 Armed Forces Qualification Test

General description – The Armed Forces Qualification Test measures mechanical and mathematical aptitude. Scores from the test are used to place Army recruits into a military occupational specialty.

Strengths and limitations – To minimize attrition due to poor placement, supplementary aptitude tests are required. Further, the results are affected by the substitutability with compensation principle. This principle states that "a relatively high ability in one area makes up for a low level in another so that observed performance equals that predicted from a linear combination of two predictor measures" (Uhlaner, 1972, p. 206).

Data requirements – Cognitive-noncognitive variance should be considered when evaluating the test scores.

Thresholds – Not stated.

Source

Uhlaner, J.E. Human performance effectiveness and the systems measurement bed. *Journal of Applied Psychology* 56(3): 202–210, 1972.

2.3.3 Deutsch and Malmborg Measurement Instrument Matrix

General description – The measurement instrument matrix consists of activities that must be performed by an organization along the vertical axis and the metrics used to evaluate the performance of those activities along the horizontal axis. A value of one is placed in every cell in which the metric is an appropriate measure of the activity, otherwise, a zero is inserted.

Strengths and limitation – This method handles the complexity and interaction of activities performed by organizations. It has been used to assess the impact of information overload on decision-making effectiveness. The measurement instrument matrix should be analyzed in conjunction with an objective matrix. The objective matrix lists the set of activities along the horizontal axis and the set of objectives along the vertical axis.

Data requirement – Reliable and valid metrics are required for each activity performed.

Threshold – Zero is the lower limit; one, the upper limit.

Source

Deutsch, S.J., and Malmborg, C.J. The design of organizational performance measures for human decision making, Part I: Description of the design methodology. IEEE Transactions on Systems, Man, and Cybernetics SMC-12(3): 344–352, 1982.

2.3.4 Performance Evaluation Tests for Environmental Research (PETER)

General description – The Performance Evaluation Tests for Environmental Research (PETER) test battery is made up of 26 tests: (1) aiming, (2) arithmetic, (3) associative memory, (4) Atari air combat maneuvering, (5) Atari anti-aircraft, (6) choice RT: 1-choice, (7) choice RT: 4-choice, (8) code substitution, (9) flexibility of closure, (10) grammatical reasoning, (11) graphic and phonemic analysis, (12) letter classification: name, (13) letter classification: category, (14) manikin, (15) Minnesota rate of manipulation, (16) pattern comparison, (17) perceptual speed, (18) search for typos in prose, (19) spoke control, (20) Sternberg item recognition: positive set 1, (21) Sternberg item recognition: positive set 4, (22) Stroop, (23) tracking: critical, (24) tracking: dual critical, (25) visual contrast sensitivity, and (26) word fluency (Kennedy, 1985). The process for test selection is described in Carter et al. (1980a). A tabular summary of the individual tests is presented in Kennedy et al. (1980).

Strengths and limitations – Tests were selected for the PETER battery on the following criteria: (1) administration time, (2) total stabilization time, and (3) reliability. Kennedy and Bittner (1978) measured performance of 19 Navy enlisted men on 10 tasks over a 15-day period. Eight tasks had significant learning effects over days (grammatical reasoning, code substitution, Stroop, arithmetic, Neisser letter search, critical tracking task, subcritical two-dimensional compensatory tracking, and Spoke test). Two did not (time estimation and complex counting). Two (time estimation and Stroop) had low reliabilities. In another study using time estimation and tracking, Bohnen and Gaillard (1994) reported that time estimation was not affected by sleep loss whereas tracking was. Seales et al. (1979) reported that performance on a paper-and-pencil arithmetic test was stable after nine days and had constant variance throughout 15 days of testing. Their participants were 18 Navy enlisted men. McCauley et al. (1979) examined time estimation performance. Their participants were 19 Navy enlisted men. Forty trials per day were performed over 15 days. There were no significant differences in performance over time. McCafferty et al. (1980) reported performance on an auditory forward digit span task was stable after four days with no significant differences in variance over 12 days. Their participants were nine Navy enlisted men. Guignard et al. (1980) reported unstable speed and error measures for the Landolt C reading test over 12 days. Their participants were eight Navy enlisted men. Carter et al. (1980b) reported that response time for the Sternberg task was stable after the fourth day in a 15-day experiment. The slope, however, was unreliable over time. The participants were 21 Navy enlisted men. Harbeson et al. (1980) reported acceptable reliability in two (interference susceptibility and free recall) tests and unacceptable reliability in two other memory tests (running recognition and list differentiation). Their participants were 23 Navy enlisted men who performed the tasks over 15 consecutive days.

Performance on individual tests (e.g., navigation plotting: Wiker et al., 1983; vertical addition, grammatical reasoning, perceptual speed, flexibility of closure, Bittner et al., 1983) were compared in the laboratory and at sea. Bittner et al. (1984) summarized the results of all 112 tests studied for possible inclusion in PETER by placing each task into one of the following categories: good, good but redundant, ugly (flawed), and bad.

Data requirements – Each PETER test has been programmed for a NEC PC 8201A.

Thresholds – Not stated.

Sources

Bittner, A.C., Carter, R.C., Kennedy, R.S., Harbeson, M.M., and Krause, M. Performance Evaluation Tests for Environmental Research (PETER): The good, bad, and ugly. Proceedings of the Human Factors Society 28th Annual Meeting, vol. 1, 11–15, 1984.

Bittner, A.C., Carter, R.C., Krause, M., Kennedy, R.S., and Harbeson, M.M. Performance Evaluation Tests for Environmental Research (PETER): Moran and computer batteries. *Aviation, Space, and Environmental Medicine* 54(10): 923–928, 1983.

Bohnen, H.G.M., and Gaillard, A.W.K. The effects of sleep loss in a combined tracking and time estimation task. *Ergonomics* 37(6): 1021–1030, 1994.

Carter, R.C., Kennedy, R.S., and Bittner, A.C. Selection of Performance Evaluation Test for Environmental Research. Proceedings of the Human Factors Society 24th Annual Meeting, 320–324, 1980a.

Carter, R.C., Kennedy, R.S., Bittner, A.C., and Krause, M. Item recognition as a Performance Evaluation Test for Environmental Research. Proceedings of the Human Factors Society 24th Annual Meeting, 340–343, 1980b.

Guignard, J.C., Bittner, A.C., Einbender, S.W., and Kennedy, R.S. Performance Evaluation Tests for Environmental Research (PETER): Landolt C reading test. Proceedings of the Human Factors Society 24th Annual Meeting, 335–339, 1980.

Harbeson, M.M., Krause, M., and Kennedy, R.S. Comparison of memory tests for environmental research. Proceedings of the Human Factors Society 24th Annual Meeting, 349–353, 1980.

Kennedy, R.S. *A Portable Battery for Objective, Non-obtrusive Measures of Human Performance (NASA-CR-171868)*. Pasadena, CA: Jet Propulsion Laboratory, 1985.

Kennedy, R.S., and Bittner, A.C. Progress in the analysis of a Performance Evaluation Test for Environmental Research (PETER). Proceedings of the Human Factors Society 22nd Annual Meeting, 29–35, 1978.

Kennedy, R.S., Carter, R.C., and Bittner, A.C. A catalogue of Performance Evaluation Tests for Environmental Research. Proceedings of the Human Factors Society 24th Annual Meeting, 344–348, 1980.

McCafferty, D.B., Bittner, A.C., and Carter, R.C. Performance Evaluation Test for Environmental Research (PETER): Auditory digit span. Proceedings of the Human Factors Society 24th Annual Meeting, 330–334, 1980.

McCauley, M.E., Kennedy, R.S., and Bittner, R.S. Development of Performance Evaluation Tests for Environmental Research (PETER): Time estimation test. Proceedings of the Human Factors Society 23rd Annual Meeting, 513–517, 1979.

Seales, D.M., Kennedy, R.S., and Bittner, A.C. Development of Performance Evaluation Test for Environmental Research (PETER): Arithmetic computation. Proceedings of the Human Factors Society 23rd Annual Meeting, 508–512, 1979.

Wiker, S.F., Kennedy, R.S., and Pepper, R.L. Development of Performance Evaluation Tests for Environmental Research (PETER): Navigation plotting. *Aviation, Space, and Environmental Medicine* 54(2): 144–149, 1983.

2.3.5 Work and Fatigue Test Battery

General description – The Simulated Work and Fatigue Test Battery was developed by the National Institute for Occupational Safety and Health to assess the effects of fatigue in the workplace. The simulated work is a data entry task. The fatigue test battery has 11 tasks: (1) grammatical reasoning, (2) digit addition, (3) time estimation, (4) simple auditory RT, (5) choice RT, (6) two-point auditory discrimination, (7) response alternation performance (tapping), (8) hand steadiness, (9) the Stanford sleepiness scale, (10) the Neuropsychiatric Research Unit (NPRU) Mood Scale adjective checklist, and (11) oral temperature. Two tasks, grammatical reasoning and simple RT, are also performed in dual-task mode.

Strengths and limitations – The Simulated Work and Fatigue Test Battery is portable, brief, easy to administer, and requires little training of the participant. Rosa and Colligan (1988) used the battery to evaluate the effect of fatigue on performance. All tasks showed significant fatigue effects except: data entry, time production, and two-point auditory discrimination.

Data requirements – The microcomputer provides all task stimuli records, as well as data, and scores all tasks.

Thresholds – Not stated.

Source

Rosa, R.R., and Colligan, M.J. Long workdays versus rest days: Assessing fatigue and alertness with a portable performance battery. *Human Factors* 30(3): 305–317, 1988.

2.3.6 Unified Tri-Services Cognitive Performance Assessment Battery (UTCPAB)

General description – The Unified Tri-services Cognitive Performance Assessment Battery (UTCPAB) is made up of 25 tests: (1) linguistic processing, (2) grammatical reasoning (traditional), (3) grammatical reasoning (symbolic), (4) two-column addition, (5) mathematical processing, (6) continuous recognition, (7) four-choice serial RT, (8) alpha-numeric visual vigilance, (9) memory search, (10) spatial processing, (11) matrix rotation, (12) manikin, (13) pattern comparison (simultaneous), (14) pattern comparison (successive), (15) visual scanning, (16) code substitution, (17) visual probability monitoring, (18) time wall, (19) interval production, (20) Stroop, (21) dichotic listening, (22) unstable tracking, (23) Sternberg-tracking combination, (24) matching to sample, and (25) item order.

Strengths and limitations – Tests were selected for the UTCPAB based on the following criteria: (1) used in at least one Department of Defense laboratory, (2) proven validity, (3) relevance to military performance, and (4) sensitivity to hostile environments and sustained operations (Perez et al., 1987).

A nine-test version, the Walter Reed Performance Assessment Battery, was used to assess the effects of altitude. Using this shortened version, Crowley et al. (1992) reported decrements on three tasks due to altitude effects: code substitution, Stroop, and logical reasoning.

Using another shortened version, the Criterion Task Set (which includes probability monitoring, unstable tracking, continuous recall, grammatical reasoning, linguistic processing, mathematical processing, memory search, spatial processing, and interval production), Chelen et al. (1993) reported no significant effects on performance of any of these tasks of phenytoin serum (motion sickness therapy) levels.

Rogers et al. (1989) used the same approach to select tests to evaluate the effects of a 1-hour nap and caffeine on performance. The tests were sustained attention, auditory vigilance and tracking, complex vigilance, two-letter cancellation, digit symbol substitution, logic, short-term memory, and visual vigilance. There were significant effects on sustained attention, auditory vigilance and tracking, visual vigilance, and complex vigilance but not on short-term memory.

Using still another task battery, Simple Portable Aviation Relevant Test battery and Answer-scoring System (SPARTANS), Stokes et al. (1994) reported no significant effects of aspartame. The tasks included in SPARTANS were: maze tracing, hidden figures recognition, hidden figures rotation, visual number, scheduling, Sternberg, first order pursuit tracking, dual task of Sternberg and tracking, minefield, and Stroop.

In another task set, Paul and Fraser (1994) reported no effect on mild acute hypoxia on the ability to learn new tasks. In this case, the tasks were Manikin, choice RT, and logical reasoning. In fact, performance on the first two tasks improved over time. In yet another task set (simple RT, four-choice RT, tracking, visual search, and visual analogue series), Cherry et al. (1983) reported decreases in performance of the tracking and visual search task as alcohol level increased. There was no effect on any of the tasks of toluene, a rubber solvent.

In still another grouping of tasks, Beh and McLaughlin (1991) investigated the effects of desynchronization on airline crews. The tasks used were: grammatical reasoning, horizontal addition, vertical addition, letter cancellation, and card sorting. There were no significant effects on the number of errors. The control group, however, completed more items on the grammatical reasoning and vertical addition tasks than the desynchronized flight crews. There was a significant group by test period interaction for the horizontal addition task and for the card sorting task.

Rosa and Bonnet (1993) applied a slightly different set to evaluate the effect of 8 and 12 hour rotating shifts on performance. Tests included mental arithmetic, dual task of grammatical reasoning and auditory RT, simple auditory RT, and hand steadiness. For mental arithmetic, there were the most correct answers on the 8-hour evening shift and fewest errors at 24 hours. For grammatical reasoning, RT the first day on an eight-hour shift was significantly longer than the fourth day, which was a 12-hour shift. There were also 9% more errors on 12-hour shifts. Dual RT was fastest at 24:00 hours. For simple RT percent misses, there were significantly more misses on the fourth day (i.e., the first day of a 12-hour shift) than any of the eight-hour shifts (day, evening, or night). Hand steadiness was 0.5% greater per day over the five-day test.

In another test battery, the Automated Performance Test System (APTS), Kennedy et al. (1993) reported eight of the nine tests showed significant effects of alcohol (up to 0.15% Blood Alcohol Concentration). The tests were preferred hand tapping, non-preferred hand tapping, grammatical reasoning, mathematical processing, code substitution, pattern comparison, manikin, short-term memory, and four-choice RT. Grammatical reasoning did not show a significant alcohol effect. Kennedy et al. (1996) compared APTS with its successor Delta. There were essentially no differences in task performance associated with hardware or software differences between the two systems.

Salame (1993) identified a major defect in the grammatical reasoning test that was part of the Standardized Tests for Research and Environmental Stressors (STRES). The defect was in the pattern of answers: if there was one match the answer was always "the same" and if there were no or two matches, the answer was always "different."

Bonnet and Arand (1994) used addition, vigilance, and logical reasoning tasks to assess the effects of naps and caffeine. The researchers concluded that naps and caffeine resulted in near baseline performance over a 24-hour period without sleep. Their participants were 18 to 30 year old males.

The utility of test batteries is demonstrated by research into the effects of shifts in work-rest cycles reported by Porcu et al. (1998). They reported no effects on a digit symbol substitution task or on the Deux Barrages task (a paper-and-pencil task requiring marking two target symbols embedded among similar symbols) but a significant decrement on a letter cancellation task.

Data requirements – Each UTCPAB test has been programmed for computer administration. Standardized instructions are used as well as a standardized data file format.

Thresholds – Not stated.

Sources

Beh, H.C., and McLaughlin, P.J. Mental performance of air crew following layovers on transzonal flights. *Ergonomics* 34(2): 123–135, 1991.

Bonnet, M.H., and Arand, D.L. The use of prophylactic naps and caffeine to maintain performance during a continuous operation. *Ergonomics* 37(6): 1009–1020, 1994.

Chelen, W., Ahmed, N., Kabrisky, M., and Rogers, S. Computerized task battery assessment of cognitive and performance effects of acute phenytoin motion sickness therapy. *Aviation, Space, and Environmental Medicine* 64(3), 201–205, 1993.

Cherry, N., Johnston, J.D., Venables, H., and Waldron, H.A. The effects of toluene and alcohol on psychomotor performance. *Ergonomics* 26(11): 1081–1087, 1983.

Crowley, J.S., Wesensten, N., Kamimori, G., Devine, J., Iwanyk, E., and Balkin, T. *Aviation, Space, and Environmental Medicine* 63(8), 696–701, 1992.

Kennedy, R.S., Dunlap, W.P., Ritters, A.D., and Chavez, L.M. Comparison of a performance test battery implements on different hardware and software: APTS versus DELTA. *Ergonomics* 39(8): 1005–1016, 1996.

Kennedy, R.S., Turnage, J.J., Wilkes, R.L., and Dunlap, W.P. Effects of graded dosages of alcohol on nine computerized repeated-measures tests. *Ergonomics* 36(10): 1195–1222, 1993.

Paul, M.A., and Fraser, W.D. Performance during mild acute hypoxia. *Aviation, Space, and Environmental Medicine* 65(10), 891–899, 1994.
Perez, W.A., Masline, P.J., Ramsey, E.G., and Urban, K.E. *Unified Tri-services Cognitive Performance Assessment Battery: Review and Methodology (AAMRL-TR-87-007).* Wright-Patterson Air Force Base, OH: Armstrong Aerospace Medical Research Laboratory, March 1987.
Porcu, S., Bellatreccia, A., Ferrara, M., and Casagrande, M. Sleepiness, alertness and performance during a laboratory simulation of an acute shift of the wake-sleep cycle. *Ergonomics* 41(8): 1192–1262, 1998.
Rogers, A.S., Spencer, M.B., Stone, B.M., and Nicholson, A.N. The influence of a 1 h nap on performance overnight. *Ergonomics* 32(10): 1193–1205, 1989.
Rosa, R.R., and Bonnet, M.H. Performance and alertness on 8 h and 12 h rotating shifts at a natural gas utility. *Ergonomics* 36(10): 1177–1193, 1993.
Salame, P. The AGARD grammatical reasoning task: A defect and proposed solutions. *Ergonomics* 36(12): 1457–1464, 1993.
Stokes, A.F., Belger, A., Banich, M.T., and Bernadine, E. Effects of alcohol and chronic aspartame ingestion upon performance in aviation relevant cognitive tasks. *Aviation, Space, and Environmental Medicine* 65, 7–15, 1994.

2.4 Domain Specific Measures

The fourth category of human performance measures is domain specific measures which assess abilities to perform a family of related tasks. These measures assume that abilities and effects vary across segments of a mission or on the use of different controllers. Examples in this category are aircraft parameters (Section 2.4.1), air traffic control performance measures (Section 2.4.2), Boyett and Conn's White-Collar Performance Measures (Section 2.4.3), Charlton's Measures of Human Performance in Space Control Systems (Section 2.4.4), driving parameters (Section 2.4.5), Eastman Kodak Company Measures for Handling Tasks (Section 2.4.6), and Haworth-Newman Avionics Display Readability Scale (Section 2.4.7).

2.4.1 Aircraft Parameters

General description – Aircrew performance is often estimated from parameters describing aircraft state. These parameters include airspeed, altitude, bank angle, descent rate, glide slope, localizer, pitch rate, roll rate, and yaw rate. Measures derived from these include root-mean-square values, minimums and maximums, correlations between two or more of these parameters, and deviations between actual and assigned values.

Strengths and limitations – One universal measure is number of errors. For example, Hardy and Parasuraman (1997) developed an error list based on component pilot activities (see Table 2.1). McGarry and Stelzer (2011) used

TABLE 2.1

Component Abilities of Commercial Airline Pilot Performance Determined by Frequency of Errors Extracted from Accident Reports, Critical Incidents, and Flight Checks

	Frequency of Errors			
Component Ability	**Accidents**	**Incidents**	**Flight Checks**	**Total**
Establishing and maintaining angle of glide, rate of descent, and gliding speed on approach to landing	47	41	11	99
Operating controls and switches	15	44	33	92
Navigating and orienting	4	39	19	62
Maintaining safe airspeed and attitude, recovering from stalls and spins	11	28	18	57
Following instrument flight procedures and observing instrument flight regulations	5	27	13	45
Carrying out cockpit procedures and routines	7	31	4	42
Establishing and maintaining alignment with runway on approach or takeoff climb	3	31	5	39
Attending, remaining alert, maintaining lookout	14	23	1	38
Utilizing and applying essential pilot information	0	19	18	37
Reading, checking, and observing instruments, dials, and gauges	1	26	7	34
Preparing and planning of flight	2	27	3	32
Judging type of landing or recovering from missed or poor landing	1	23	8	32
Breaking angle of glide on landing	1	25	5	31
Obtaining and utilizing instructions and information from control personnel	3	21	0	24
Reacting in an organized manner to unusual or emergency situations	0	17	7	24
Operating plane safely on ground	7	15	1	23
Flying with precision and accuracy	0	7	15	22
Operating and attending to radio	0	7	10	17
Handling of controls smoothly and with coordination	0	6	8	14
Preventing plane from undue stress	0	5	7	12
Taking safety precautions	2	5	4	11

Hardy, D.J., and Parasuraman, R. Cognition and flight performance in older pilots. *Journal of Experimental Psychology* 3(4): 313–348, 1997.

errors to investigate the concurrent use of Runway Entrance Lights (RELs) and Surface Movement Guidance Control System (SMGCS) stop bars during taxi in low visibility conditions. They reported none of the trained pilots failed to stop while one of the untrained pilots did.

Not all universal aircraft parameters are sensitive to nonflying stressors. For example, Wierwille et al. (1985) reported that neither pitch nor roll high-pass mean square scores from a primary simulated flight task were sensitive to changes in the difficulty of a secondary mathematical problem-solving task.

Researchers have also examined aircraft parameters sensitive to phase of flight (takeoff, climb, cruise, approach, and landing) and task (air combat, hover, standard rate turn). Also applied are measures of control input activity as well as composite scores. In an unusual study, Casto and Casali (2010) evaluated participants who were either non-hearing waivered or hearing waivered. The researchers also manipulated workload by varying visibility, number of maneuvers, and amount of information presented during communications. Heading, altitude, and airspeed deviations increased as workload increased as well as the number of requests for Air Traffic Control to repeat a directive.

Sources

Casto, K.L., and Casali, J.G. Effect of communications headset, hearing ability, flight workload, and communications signal quality on pilot performance in an Army Black Hawk helicopter simulator. Proceedings of the Human Factors and Ergonomics Society 54th Annual Meeting, 80–84, 2010.

Hardy, D.J., and Parasuraman, R. Cognition and flight performance in older pilots. *Journal of Experimental Psychology* 3(4): 313–348, 1997.

McGarry, K., and Stelzer, E. An assessment of pilots' concurrent use of runway entrance lights and surface movement control system guidance system stop bars. Proceedings of the Human Factors and Ergonomics Society 55th Annual Meeting, 31–35, 2011.

Wierwille, W.W., Rahimi, M., and Casali, J.G. Evaluation of 16 measures of mental workload using a simulated flight task emphasizing mediational activity. *Human Factors* 27(5): 489–502, 1985.

2.4.1.1 Takeoff and Climb Measures

General description – Airspeed and pitch are typically used as measures during aircraft takeoff and climb.

Strengths and limitations – Cohen (1977), using a flight simulator, reported increased airspeed within the first 60 seconds after being launched from the deck of an aircraft carrier. Airspeed was also significantly different among three types of flight displays as were variance of airspeed, vertical speed, altitude, angle of attack, pitch attitude, and frequency of pitch adjustments.

Williams (2000) measured horizontal and vertical deviations from the pathway during climb. He reported significantly smaller errors for both the horizontal and vertical when acquiring a pathway using a highway-in-the-sky display than with a follow-me aircraft display. The participants were 36 pilots flying a general aviation simulator.

In an evaluation of Unmanned Aerial System (UAS) performance, Fern et al. (2012) used the minimum horizontal and vertical separation distances between the UAS and other aircraft as well as the number of losses of separation (vertical separation less than 750 feet and horizontal separation less than 5 nm) as the dependent measures to compare performance with or without a Cockpit Situation Display and in high or low traffic density. There was no significant effect of display but there were significantly more losses of separation in high traffic density compared to low traffic density.

Data requirements – The simulator and/or aircraft must be instrumented.

Thresholds – Not stated.

Sources

Cohen, M.M. Disorienting effects of aircraft catapult launching: III. Cockpit displays and piloting performance. *Aviation, Space, and Environmental Medicine* 48(9): 797–804, 1977.

Fern, L., Kenny, C.A., Shively, R.J., and Johnson, W. UAS integration into the NAS: An examination of baseline compliance in the current airspace system. Proceedings of the Human Factors and Ergonomics Society 56th Annual Meeting, 41–45, 2012.

Williams, K.W. Age and situation awareness: A highway-in-the-sky display study. In D.B. Kaber and M.R. Endsley (Eds.) Proceedings of the First Human Performance, Situation Awareness and Automation: User-Centered Design for the New Millennium, October 15–19, 2000.

2.4.1.2 Cruise Measures

In an in-flight helicopter study, Berger (1977) reported significant differences in airspeed, altitude, bank angle, descent rate, glide slope, localizer, pitch rate, roll rate, and yaw rate and some combination of Visual Meteorological Conditions (VMC), Instrument Meteorological Conditions (IMC) with fixed sensor, IMC with stabilization sensor, and IMC with a sensor looking ahead through turns.

North et al. (1979) reported that: (1) pitch error, heading error, roll acceleration, pitch acceleration, speed error, and yaw position were sensitive to differences in display configurations, (2) pitch error, roll error, heading error, roll acceleration, pitch acceleration, yaw acceleration, speed error, pitch position, roll position, yaw position, power setting, altitude error, and cross track error were sensitive to differences in winds, and (3) heading error, roll acceleration, pitch acceleration, yaw acceleration, speed error, roll position, yaw position, altitude error, and cross track error were sensitive to motion cues.

Bortolussi and Vidulich (1991) calculated a figure of merit (FOM) for simulated flight from rudder standard deviation (SD), elevator SD, aileron SD, altitude SD, mean altitude, airspeed SD, mean airspeed, heading SD, and mean heading. Only the airspeed and altitude FOMs showed significant differences between scenarios. The primary measures for these variables were also significant as well as aileron SD and elevator SD.

In an evaluation of three-dimensional sonifications to present aircraft waypoint bearing and course deviation, Towers et al. (2014) reported that the auditory display significantly improved heading and course deviation accuracy as well as head up time.

Janowsky et al. (1976) reported significant decrements in the following performance measures after pilots smoked 0.9 mg/kg of marijuana: number of major and minor errors, altitude deviations, heading deviations, and radio navigation errors. The data were collected from 10 male pilots who had smoked marijuana socially. The aircraft simulator was a general aviation model used for instrument flight training.

Dattel et al. (2015) compared performance of pilots who received procedural training in traffic patterns or conceptual training or practice sessions in a flight simulator. There were significant effects of training group on distance deviation with the conceptual training group having the smallest deviations. There were, however, no significant effects of training on altitude deviation, heading deviation, or airspeed deviation.

In an airplane upset recovery experiment, Landman et al. (2017) compared the adherence to recovery procedures when airline pilots anticipated a stall event during cruise and when they did not. The data were collected in a flight simulator. The pilots had greater adherence to the procedures when the event was anticipated than when it was not.

Data requirements – The simulator and/or aircraft must be instrumented.

Thresholds – Not stated.

Sources

Berger, I.R. Flight performance and pilot workload in helicopter flight under simulated IMC employing a forward looking sensor. Proceedings of Guidance and Control Design Considerations for Low-Altitude and Terminal-Area Flight (AGARD-CP-240). AGARD, Neuilly-sur-Seine, France, 1977.

Bortolussi, M.R., and Vidulich, M.A. An evaluation of strategic behaviors in a high fidelity simulated flight task. Comparing primary performance to a figure of merit. Proceedings of the 6th International Symposium on Aviation Psychology, vol. 2, 1101–1106, 1991.

Dattel, A.R., Karunratanakul, K., Crockett, S.A., and Fabbri, J. How procedural and conceptual training affect flight performance for learning traffic patterns. Proceedings of the Human Factors and Ergonomics Society 59th Annual Meeting, 855–858, 2015.
Janowsky, D.S., Meacham, M.P., Blaine, J.D., Schoor, M., and Bozzetti, L.P. Simulated flying performance after marihuana intoxication. *Aviation, Space, and Environmental Medicine* 47(2): 124–128, 1976.
Landman, A., Groen, E.L., van Paassen, M.M., Bronkhorst, A.W., and Mulder, M. The influence of surprise on upset recovery performance in airline pilots. *International Journal of Aerospace Psychology* 27(1–2): 2–14, 2017.
North, R.A., Stackhouse, S.P., and Graffunder, K. *Performance, Physiological, and Oculometer Evaluation of VTOL Landing Displays (NASA-CR-3171).* Hampton, VA: NASA Langley Research Center, 1979.
Towers, J., Burgess-Limerick, R., and Riek, S. Concurrent 3-D sonifications enable the head-up monitoring of two interrelated aircraft navigation instruments. *Human Factors* 56(8): 1414–1427, 2014.

2.4.1.3 Approach and Landing Measures

Morello (1977), using a B-737 aircraft, reported differences in localizer, lateral, and glide slope deviations during three-nautical-mile and close-in approaches between a baseline and an integrated display format. However, neither localizer rmse nor glide slope rmse were sensitive to variations in pitch-stability level, wind-gust disturbance, or crosswind direction and velocity (Wierwille and Connor, 1983).

Brictson (1969) reported differences between night and day carrier landings for the following aircraft parameters: altitude error from glide slope, bolster rate, wire arrestments, percent of unsuccessful approaches, and probability of successful recovery. There was no significant day–night difference for lateral error from centerline, sink speed, and final approach airspeed.

Kraft and Elworth (1969) reported significant differences in generated altitude during final approach due to individual differences, city slope, and lighting. Also related to individual differences, Billings et al. (1975) compared the performance of five experienced pilots in a Cessna 172 and a Link Singer General Aviation Trainer during Instrument Landing System (ILS) approaches in IMC. Pitch, roll, and airspeed errors were twice as large in the aircraft as in the simulator. Further, there were improvements in performance in the simulator over time but not in the aircraft.

Swaroop and Ashworth (1978) reported significantly higher glide slope intercepts and flight path elevation angles with a diamond marking on the runway than without a diamond marking on the runway. Touchdown distance was significantly smaller without the diamond for research pilots and with the diamond for general aviation pilots. Also examining the effect of displays, Lewis and Mertens (1979) reported significant differences in rmse deviation on glide path approach angle among four different displays.

In another display study, Lintern et al. (1984) reported significant differences between conventional and modified Fresnel Lens Optical Landing System displays for glide slope rmse and descent rate at touchdown but not for localizer rmse. From the same research facility, Lintern and Koonce (1991) reported significant differences in vertical angular glide slope errors among varying scenes, display magnifications, runway sizes, and start points.

In a comparison of quickened and nonquickened displays, Spielman et al. (2014) reported significant differences in the standard deviation and rmse of altitude error during approach and landing. There were no significant effects on lateral errors, however. Their participants were not pilots.

Examining an infrequent but critical maneuver, Dehais et al. (2017) examined errors during a go around in a simulator. Only three out of 12 crews were able to read back the complete go-around clearance. Six crews made trajectory errors. Other errors included failure to retract landing gear, dialing an erroneous heading, not correctly engaging heading mode, not setting altitude in flight control unit, setting altitude in late, and not engaging altitude properly.

Instead of relying on individual measures, Gaidai and Mel'nikov (1985) developed an integral criterion to evaluate landing performance:

$$I(t) = \frac{1}{t_z} \int_0^{t_z} \sum_{j=1}^{K} a_i(t_i) \left[\frac{Y_i(t_i) - m_{yi}}{s_{yi}} \right]^2 dt$$

where

I = integral criteria

t = time

t_z = integration time

$Ka_j(t_i)$ = weighting coefficient for the path parameter Y_i at instant i

$Y_j(t_i)$ = instantaneous value of parameter Y_j at instant i

m_{Yj} = programmed value of the path parameter

s_{Yj} = standard deviation for the integral deviations:

$$s_{y_j} = \frac{1}{n_{j=1}} n \sqrt{\frac{1}{t_z} \int_0^{t_z} \left[Y_j(t_i) - m_{yi} \right]^2 dt}$$

$I(t)$ provides a multivariate assessment of pilot performance but is difficult to calculate.

Also using a set of measures, Bhagat et al. (2011) compared the Flight Safety Foundation (FSF) Approach and Landing Accident Reduction (ALAR) Toolkit safety profile to the operational limits for approach and landing. The ALAR measures were height over threshold, final approach speeds, touchdown distance, deceleration rate, and runway exit speeds.

Observational data were provided by a major airline for a 14-month period. There were exceedances from the standard operating procedure values for all variables.

With the increased amount of automation in the cockpit, pilots' ability to manually fly aircraft has been questioned. In a simulator study with 57 airline pilots, Haslbeck et al. (2014) reported significantly larger maximum localizer and glideslope deviations as well as rmse error on each for Captains than for First Officers.

Another approach using multiple measures is the Landing Performance Score (LPS), which is a score derived from the multiple regression of the following variables: number of landings per pilot, log book scorings, environmental data (weather, sea state, etc.), aircraft data (type and configuration), carrier data (ship size, visual landing aids, accident rate, etc.), boarding and bolster rate, intervals between landings, mission type and duration, and flying cycle workload estimate. LPS was developed for Navy carrier landings (Brictson, 1977). LPS distinguished night and day carrier landings (Brictson, 1974).

In a UAV study, Draper et al. (2000) used absolute landing error from touchdown marker to assess the effects of type of alter, visibility level, and turbulence severity. The data were collected in a simulated UAV ground station from eight rated pilots. There was significantly less error with a haptic cue compared to without a haptic cue.

Sources

Bhagat, R., Glussick, D., Histon, J., and Saccomanno, F. Formulating safety performance measures for aircraft landing and runway exit maneuvers. Proceedings of the Human Factors and Ergonomics Society 55th Annual Meeting, 1725–1729, 2011.

Billings, C.E., Gerke, R.J., and Wick, R.L. Comparisons of pilot performance in simulated and actual flight. *Aviation, Space, and Environmental Medicine* 46(3): 304–308, 1975.

Brictson, C.A. Operational measures of pilot performance during final approach to carrier landing. Proceedings of Measurement of Aircrew Performance – The Flight Deck Workload and Its Relation to Pilot Performance (AGARD-CP-56). AGARD, Neuilly-sur-Seine, France, 1969.

Brictson, C.A. Pilot landing performance under high workload conditions. In A.N. Nickolson (Ed.) *Simulation and Study of High Workload Operations (AGARD-CP-146)*. Neuilly-sur-Seine, France: AGARD, 1974.

Brictson, C.A. Methods to assess pilot workload and other temporal indicators of pilot performances effectiveness. In R. Auffret (Ed.) Advisory Group for Aerospace Research and Development (AGARD) (Conference Proceedings Number 217, AGARD-CP-217), B9-7–B9-10, 1977.

Dehais, F., Behrend, J., Peysakhovich, V., Causse, M., and Wickens, C.D. Pilot flying and pilot monitoring's aircraft state awareness during go-around execution in aviation: A behavioral and eye tracking study. *International Journal of Aerospace Psychology* 27(1–2): 15–28, 2017.

Draper, M.H., Ruff, H.A., Repperger, D.W., and Lu, L.G. Multi-sensory interface concepts supporting turbulence detection by UAV controllers. In D.B. Kaber and M.R. Endsley (Eds.) Proceedings of the First Human Performance, Situation Awareness and Automation: User-Centered Design for the New Millennium, October 15–19, 2000.

Gaidai, B.V., and Mel'nikov, E.V. Choosing an objective criterion for piloting performance in research on pilot training on aircraft and simulators. *Cybernetics and Computing Technology* 3: 162–169, 1985.

Haslbeck, A., Kirchner, P., Schubert, E., and Bengler, K. A flight simulator study to evaluate manual flying skills of airline pilots. Proceedings of the Human Factors and Ergonomics Society 58th Annual Meeting, 11–15, 2014.

Kraft, C.L., and Elworth, C.L. Flight deck work and night visual approach. Proceedings of Measurement of Aircrew Performance – The Flight Deck Workload and its Relation to Pilot Performance (AGARD-CP-56). AGARD, Neuilly-sur-Seine, France, 1969.

Lewis, M.F., and Mertens, H.W. Pilot performance during simulated approaches and landings made with various computer-generated visual glidepath indicators. *Aviation, Space, and Environmental Medicine* 50(10): 991–1002, 1979.

Lintern, G., Kaul, C.E., and Collyer, S.C. Glide slope descent-rate cuing to aid carrier landings. *Human Factors* 26(6): 667–675, 1984.

Lintern, G., and Koonce, J.M. Display magnification for simulated landing approaches. *International Journal of Aviation Psychology* 1(1): 59–72, 1991.

Morello, S.A. Recent flight test results using an electronic display format on the NASA B-737. Proceedings of Guidance and Control Design Considerations for Low-Altitude and Terminal Area Flight (AGARD-CP-240). AGARD, Neuilly-sur-Seine, France, 1977.

Spielman, Z.A., Evans, R.T., Holmberg, J.D., and Dyre, B.P. Evaluation of a peripherally-located instrument landing display with high-order control of a nonlinear approach and landing. Proceedings of the Human Factors and Ergonomics Society 58th Annual Meeting, 1067–1071, 2014.

Swaroop, R., and Ashworth, G.R. *An Analysis of Flight Data from Aircraft Landings with and without the Aid of a Painted Diamond on the Same Runway (NASA-CR-143849)*. Edwards Air Force Base, CA: NASA Dryden Research Center, 1978.

Wierwille, W.W., and Connor, S.A. Sensitivity of twenty measures of pilot mental workload in a simulated ILS task. Proceedings of the Annual Conference on Manual Control (18th), 150–162, 1983.

2.4.1.4 Air Combat Measures

Kelly (1988) reviewed approaches to automated aircrew performance measurement during air-to-air combat. He concluded that measures must include positional advantage or disadvantage, control manipulation, and management of kinetic and potential energy.

Barfield et al. (1995) examined the effect of frame of reference, field of view (FOV), and eye point elevation on performance of a simulated air-to-ground targeting task. There was a significant frame of reference effect. Specifically, the pilot's eye display was associated with lower flight-path rmse, faster target lock-on time, and faster target acquisition time than the God's eye display. There was also a significant effect of FOV: lower rmse for 30 degrees and 90 degrees FOV than for 60 degrees FOV, fastest time to lock out target for 30 degrees FOV, and fastest target acquisition time. Finally, eye point elevation also resulted in significant differences: lower rmse, faster lock-on times, and faster target acquisition times for 60 degrees than 30 degrees elevation.

Kruk et al. (1983) compared the performance of 12 experienced fighter pilots, 12 undergraduate training instructor pilots, and 12 student pilots on three tasks (formation, low level, and landing) in a ground simulator. Only one parameter was sensitive to flight experience – students were significantly poorer than instructors in the distance of first correction to the runway during the landing task. The measures were: time spent in correct formation, percentage of bomb strikes within 36 m of target center, time in missile tracking, number of times shot down, number of crashes, altitude variability, heading variability, release height variability, and gravity load at release.

Sources

Barfield, W., Rosenberg, C., and Furness, T.A. Situation awareness as a function of frame of reference, computer-graphics eye point elevation, and geometric field of view. *The International Journal of Aviation Psychology* 5(3): 233–256, 1995.

Kelly, M.J. Performance measurement during simulated air-to-air combat. *Human Factors* 30(4): 495–506, 1988.

Kruk, R., Regan, D., and Beverley, K.I. Flying performance on the advanced simulator for pilot training and laboratory tests of vision. *Human Factors* 25(4): 457–466, 1983.

2.4.1.5 Hover Measures

General description – Hover is a maneuver performed in a helicopter or Vertical Takeoff and Landing (VTOL) aircraft in which the aircraft holds a constant position suspended in the air.

Strengths and limitations – Moreland and Barnes (1969) derived a measure of helicopter pilot performance using the following equation:

100 − (absolute airspeed error + absolute altitude error + absolute heading error + absolute change in torque).

This measure decreased when cockpit temperature increased above 85 degrees F, was better in light to moderate than no turbulence, and was sensitive to basic piloting techniques. However, it was not affected by either clothing or equipment configurations.

Richard and Parrish (1984) used a vector combination of errors (VCE) to estimate hover performance. VCE was calculated as follows:

$$\text{VCE} = \left(x^2 + y^2 + z^2\right)^{1/2}$$

where:

x, y, and z refer to the x, y, and z axis errors.

The authors argue that VCE is a good summary measure since it discriminates trends in the data.

Data requirements – Not stated.

Thresholds – Not stated.

Sources

Moreland, S., and Barnes, J.A. Exploratory study of pilot performance during high ambient temperatures/humidity. Proceedings of Measurement of Aircrew Performance (AGARD-CP-56). AGARD, Neuilly-sur-Seine, France, 1969.

Richard, G.L., and Parrish, R.V. Pilot differences and motion cuing effects on simulated helicopter hover. *Human Factors* 26(3): 249–256, 1984.

2.4.1.6 Standard Rate Turn

General description – A Standard Rate Turn (SRT) is a turn of three degrees compass heading per second. It is taught in both rotary and fixed wing aircraft.

Strengths and limitations – Chapman et al. (2001) used mean and standard deviations from assigned altitude, airspeed, and heading to evaluate direction of turn (left versus right), degree of turn (180 versus 360), and segment (roll in, roll out, first 30, last 30). There were significant main effects for all three independent variables. Degree of turn error and altitude error were significantly greater for right than left turns and for 360 versus 180 turns. Altitude error was greater for right than left turns. Airspeed error was greater for 180 than for 360 degree turns and for the first 30 segments than for roll out. Degree of turn error standard deviation was significantly greater for roll out than roll in. Altitude error standard deviation was significantly greater for left than right turns, 180 than 360 degree turns, and first 30 than last 30 segments. Airspeed error standard deviation was significantly greater for left than right turns and for 180 than for 360 degree turns.

Data requirements – The variables are aircraft- and mission-specific.

Thresholds – Stated for TH-57 helicopter.

Source

Chapman, F., Temme, L.A., and Still, D.L. The performance of the standard rate turn (SRT) by student Naval helicopter pilots. *Aviation, Space, and Environmental Medicine* 72(4): 343–351, 2001.

2.4.1.7 Control Input Activity

General description – Corwin et al. (1989) used control input activity for the wheel (aileron) and column (elevator) as a measure of flight-path control. Griffith et al. (1984) defined control reversal rate as "the total number of control reversals in each controller axis divided by the interval elapsed time" (p. 993). They computed separate control reversal rates for steady state and maneuver transition intervals in a ground-based flight simulator.

Strengths and limitations – Corwin et al. (1989) reported that control input activity was a reliable and valid measure. Griffith et al. (1984) were unable to compute control reversal rates for throttle and rudder pedal activity due to minimal control inputs. Further, there were no significant differences among display configurations for pitch-axis control reversal rate. There were, however, significant differences among the same display configurations in the roll-axis control reversal rate.

Wierwille et al. (1985) used the total number of elevator, aileron, and rudder inputs during simulated flight. This measure was not sensitive to difficulty of a mathematical problem-solving task performed during simulated flight.

Other control input measures are the correctness and the latency of the response to emergency situations. Loveday and Wiggins (2014) reported significantly faster responses to emergencies when icons showing the cause of the failure (transponder, heading, or engine) were used in the simulator than when generic or instrument icons were used.

Data requirements – Both control reversals and time must be simultaneously recorded.

Thresholds – Not available.

Sources

Corwin, W.H., Sandry-Garza, D.L., Biferno, M.H., Boucek, G.P., Logan, A.L., Jonsson, J.E., and Metalis, S.A. *Assessment of Crew Workload Measurement Methods, Techniques and Procedures. Volume 1 – Process, Methods, and Results (WRDC-TR-89-7006).* Wright-Patterson Air Force Base, OH, 1989.

Griffith, P.W., Gros, P.S., and Uphaus, J.A. Evaluation of pilot performance and workload as a function of input data rate and update frame rate on a dot-matrix graphics display. Proceedings of the National Aerospace and Electronics Conference, 988–995, 1984.

Loveday, T., and Wiggins, M.W. Using iconic cues to recover from fixation on tablet devices in the cockpit. Proceedings of the Human Factors and Ergonomics Society 58th Annual Meeting, 350–354, 2014.

Wierwille, W.W., Rahimi, M., and Casali, J.G. Evaluation of 16 measures of mental workload using a simulated flight task emphasizing mediational activity. *Human Factors* 27(5): 489–502, 1985.

2.4.1.8 Composite Scores

General description – Many composite scores have been developed. One of the first was the Pilot Performance Index. Based on input from subject matter experts, Stein (1984) developed a list of performance variables and associated performance criteria for an air transport mission. The list was subsequently reduced by eliminating those performance measures that did not distinguish experienced from novice pilots (see Table 2.2). This collection of performance measures was called the Pilot Performance Index (PPI). Several similar composite scores exist and are described in the strengths and limitations section.

The Swedish Defense Research Agency (FOI) Pilot Performance Scale (PPS) is a questionnaire that has been used extensively in training simulators and after flights (Berggren, 2000). It has six dimensions: Operative Performance, Situational Awareness, Pilot Mental Workload, Mental Capacity, Information Handling Tactical Situation Display, and Information Handling Tactical Information Display (Smith et al., 2007). It takes about five minutes to complete.

Strengths and limitations – The PPI provides objective estimates of performance and can distinguish experienced and novice pilots. It does not measure the amount of effort being applied by the pilot, however. In an early study, Simmonds (1960) compared the performance of 17 pilots with varying levels of experience on turning while changing altitude tasks. Simmonds reported that the effect of experience was greater on consistency of performance than on accuracy of performance.

Stave (1979) used deviations from assigned flight parameters to assess the effects of vibration on pilot performance. The data were collected in a helicopter simulator. Measures included navigation error (average distance

TABLE 2.2

Pilot Performance Index Variable List

Takeoff	Initial Approach
Pitch Angle	Heading
	Manifold Left
Climb	Manifold Right
Heading	Bank Angle
Airspeed	
	Final Approach
Enroute	Heading
Altitude	Gear Position
Pitch Angle	Flap Position
Heading	Course Deviation Indicator
Course Deviation Indicator	
Omni Bearing Sensor	
Descent	
Heading	
Airspeed	
Bank Angle	
Course Deviation Indicator	
Omni Bearing Sensor	

From Stein (1984, p. 20).

off course from takeoff to ILS), ILS score (deviation from glide slope altitude, degrees of heading, and airspeed), hover score (average altitude error plus average distance off course), hover time (duration from crossing load marker to load/unload), and load position (distance between load and helicopter when tensioned or released). There was a tendency for performance to improve with increased vibration which was interpreted as compensation by the pilots.

Leirer et al. (1989) also used deviations from an ideal flight as a measure of pilot performance. They reported decrements in a fixed-base flight simulator in older pilots (18 to 29 versus 30 to 48 years old), in turbulence, and after ingestion of marijuana. In a follow-on study, Leirer et al. (1991) reported significant effects of marijuana on pilot performance 24 hours after dosing. In another drug study, Izraeli et al. (1990) used deviations in airspeed, true heading, altitude, vertical velocity and bank to evaluate the effects of pyridostigmine bromide. They reported no significant effects from a 30-mg dose. In yet another drug study, Caldwell et al. (1992) reported significant effects of 4 mg of atropine sulfate on pilot performance in a helicopter simulator. Significant deviations occurred in the following flight parameters: heading, vertical ascent heading during hover, and airspeed control. There were significant interactive effects of atropine with flight and maneuver on vertical speed control and of atropine with flight on bank angle. Ross et al. (1992) also used pilot performance deviations to examine the effects of a drug, but the drug was alcohol. They reported significant effects but only during high workload conditions (e.g., turbulence, cross wind, wind shear).

Pilot performance measures included aircraft control errors (e.g., deviation from assigned altitude, heading, and bank), navigation errors, and communication errors. In a similar study, Morrow et al. (1993) reported reduced performance associated with 0.10% Blood Alcohol Level (BAL) on pilot performance in a flight simulator. These authors used a summary of the deviations from ideal performance as the dependent variable.

Wildzunas et al. (1996) examined the effect of time delay on pilot performance in a full motion flight simulator. They calculated a composite performance score to reflect the degree to which the participant maintained the required standards for all maneuvers (takeoff, standard rate turns, straight and level flight, descending turn, approach, hover, hover turn, nap-of-the-earth flight, formation flight, and pinnacle landing). They reported significant decrements associated with the 400 and 533 ms visual display.

Paul (1996) had Canadian Forces pilots perform an instrumented departure, turn, and instrument landing in a simulator after G-Loss of Consciousness (G-LOC) induced in a centrifuge. They calculated root mean square error scores on 11 flight parameters – again using deviations from target values. Only 1 of the 29 pilots had a significant decrement in performance after G-LOC.

Reardon et al. (1998) examined flight performance in a helicopter simulator while nine military pilots were exposed to either 70-degree Fahrenheit or 100-degree Fahrenheit temperatures while wearing standard or chemical protective uniforms. Composite scores were derived from deviations from assigned aircraft states during hover, hover turn, straight and level, left climbing turn, left descending turn, right standard rate turn, contour, and nap-of-the-earth flight. There was a significant decrement in performance while wearing the chemical protective suit.

McClernon et al. (2011) calculated standardized variability scores for pitch, roll, pitch rate, roll rate, lateral acceleration, longitudinal acceleration, and normal acceleration using Z = (Vp – Mean Vs)/SDVs. Vp was a participant's variability score, mean Vs was the mean across all participants, and SDV was the standard deviation across all participants. There was a significant effect of stress training on performance in a Piper Archer aircraft. The 30 participants had no previous flight experience and received either stress training (application of cold pressor) or not.

Johnson and Wiegmann (2011) measured performance of 16 pilots flying a simulated cross-country flight in marginal weather. Performance was measured as (Total Time – (Illegal Time + 0.55 scud running time violating one weather parameter for VFR + scud running time violating two weather parameters for VFR))/total time. They reported a significant correlation of performance and amount of IMC time.

The FOI PPS has been associated with high reliability (0.73–0.90) and is significantly correlated with the NASA TLX (+0.84) and the Bedford Rating Scale (+0.69) (Smith et al., 2000).

Data requirements – The variables are aircraft- and mission-specific.

Thresholds – Not stated for any of the composite scores.

Sources

Berggren, P. *Situational Awareness, Mental Workload, and Pilot Performance – Relationships and Conceptual Aspects (FOA-R-00-01438-706-SE).* Linköping: Human Sciences Division, 2000.

Caldwell, J.A., Stephens, R.L., Carter, D.J., and Jones, H.D. Effects of 2 mg and 4 mg atropine sulfate on the performance of U.S. Army helicopter pilots. *Aviation, Space, and Environmental Medicine* 63(10): 857–864, 1992.

Izraeli, S., Avgar, D., Almog, S., Shochat, I., Tochner, Z., Tamir, A., and Ribak, J. The effect of repeated doses of 30 mg pyridostigmine bromide on pilot performance in an A-4 flight simulator. *Aviation, Space, and Environmental Medicine* 61(5): 430–432, 1990.

Johnson, C.M., and Wiegmann, D.A. Pilot error during visual flight into instrument weather: An experiment using advanced simulation and analysis methods. Proceedings of the Human Factors and Ergonomics Society 55th Annual Meeting, 138–142, 2011.

Leirer, V.O., Yesavage, J.A., and Morrow, D.G. Marijuana, aging, and task difficulty effects on pilot performance. *Aviation, Space, and Environmental Medicine* 60(12): 1145–1152, 1989.

Leirer, V.O., Yesavage, J.A., and Morrow, D.G. Marijuana carry-over effects on aircraft pilot performance. *Aviation, Space, and Environmental Medicine* 62(3): 221–227, 1991.

McClernon, C.K., McCauley, M.E., O'Connor, P.E., and Warm, J.S. Stress training improves performance during a stressful flight. *Human Factors* 53(3): 207–218, 2011.

Morrow, D., Yesavage, J., Leirer, V., Dohlert, N., Taylor, J., and Tinklenberg, J. The time-course of alcohol impairment of general aviation pilot performance in a Frasca 141 Simulator. *Aviation, Space, and Environmental Medicine* 64(8): 697–705, 1993.

Paul, M.A. Instrument flying performance after G-induced loss of consciousness. *Aviation, Space, and Environmental Medicine* 67(11): 1028–1033, 1996.

Reardon, M.J., Fraser, E.B., and Omer, J.M. Flight performance effects of thermal stress and two aviator uniforms in a UH-60 helicopter simulator. *Aviation, Space, and Environmental Medicine* 69(9): 569–576, 1998.

Ross, L.E., Yeazel, L.M., and Chau, A.W. Pilot performance with blood alcohol concentrations below 0.04%. *Aviation, Space, and Environmental Medicine* 63(11): 951–956, 1992.

Simmonds, D.C.V. An investigation of pilot skill in an instrument flying task. *Ergonomics* 3(3): 249–254, 1960.

Smith, E., Borgvall, J., and Lif, P. *Team and Collective Performance Measurement*. Bedford Technology Park Thurleigh, Bedfordshire: Policy and Capability Studies, 1 July 2007.

Stave, A.M. The influence of low frequency vibration on pilot performance (as measured in a fixed base simulator). *Ergonomics* 22(7): 823–835, 1979.

Stein, E.S. The Measurement of Pilot Performance: A Master-Journeyman Approach (DOT/FAA/CT-83/15). Atlantic City, NJ: Federal Aviation Administration Technical Center, May 1984.

Wildzunas, R.M., Barron, T.L., and Wiley, R.W. Visual display delay effects on pilot performance. *Aviation, Space, and Environmental Medicine* 67(3): 214–221, 1996.

2.4.2 Air Traffic Control Performance Measures

General description – Rantanen (2004) reviewed performance measures that have been applied in Air Traffic Control (ATC). The first was direct observation or Over-The-Shoulder (OTS) method. The second was a set of subjective measures in which participants rated their own performance and workload. His final set consisted of 40 objective measures (see Table 2.3) derived from the Performance and Objective Workload Evaluation Research (POWER) analysis of the Data Analysis and Reduction Tool (DART) and the National Track Analysis Program (NTAP).

Strengths and limitations – Rantanen (2004) listed the disadvantages of the OTS method as being labor intensive, time consuming, and potentially inaccurate. Rantanen (2004) concluded that subjective measures are easy to use and inexpensive but may be intrusive or if completed after performance inaccurate. The Rantanen (2004) measures include both Air Traffic Controller communications and resultant aircraft state. Some researchers have focused on one or the other while other researchers have used other combinations.

Communications Hah et al. (2010) used the number of Push-To-Talk voice transmissions to evaluate the effects of Data Communication (Data Comm) on controller activity. There were no significant differences among the four human machine interfaces (graphic, keyboard, template, combined). There were differences, however, in the number of Push-To-Talk transmissions associated with percent equipage (0, 10, 50, or 100%) in the duration of the transmissions.

Sanchez and Smith (2010) used response times of controllers to runway safety alerts for incursions on the takeoff runway, on the takeoff runway while the controller was providing heading information to an aircraft on queue for departure, on the takeoff runway with an unauthorized crossing, and on the arrival runway. There were no significant differences among scenarios.

TABLE 2.3

Air Traffic Controller Performance Measures (Rantanen, 2004)

Control Duration
Correlation Time
Parabolic Stability Parameter
Response Latency
Traffic Count
Transition Slope – slope of traffic increase over time
Transition Time – time from light traffic to heavy traffic
Variability in Aircraft Altitudes
Variability in Aircraft Headings
Variability in Aircraft Speeds
Violation of Separation Minima

Askelson et al. (2013) used response times to ATC commands to compare unmanned and manned aircraft.

Vu et al. (2014) used the number of communication step ons between an Air Traffic Controller and a UAS operator to identify a maximum acceptable communication latency.

Aircraft State Two of the most critical measures of Air Traffic Controller performance are conflict and error. Hadley et al. (1999) proposed standardized definitions for the following air traffic controller performance measures:

> *Conflict: Violation of safe separation minima between two aircraft. In terminal airspace, a conflict occurs when the distance between two aircraft is <3 miles laterally and <1000 ft vertically. En route conflicts occur when spacing becomes <5 miles laterally and <1000 ft vertically. At altitude above Flight Level 290, the minimum vertical separation distance is 2000 ft (FAA, 1998). There are exceptions, such as when one pilot sees the aircraft ahead and accepts visual separation, or both aircraft are established on parallel localizers. (p. 7)*
>
> *Error (Conflict and Non-conflict): A conflict error (operational error) occurs when a failure of equipment, human, procedural, and/or system elements, individually or in combination, results in less than the separation minima. Non-conflict errors include, but are not limited to, misidentification of information from the radar display, acceptance of incomplete position information, and interpreting flight progress strips incorrectly. (p. 7)*

Rovira and Parasuraman (2010) used conflict detection and resolution performance to evaluate the effects of conflict probe technology in a mixed equipage fleet. Conflicts were detected significantly faster and more accurately with reliable automation rather than manual performance.

Khadilkar and Balakrishnan (2013) used the following to evaluate Boston Logan airport operations: time spent in departure queue as a function of queue length when the aircraft enters it, departure throughput, taxi-out time, and runway utilization.

Strybel et al. (2016) used numbers of loss of separation and spacing efficiency to compare the performance of retired Air Traffic Controllers using four different separation assurance and spacing concepts in enroute and transitional sectors. The authors reported that the numbers of loss of separation was highest when controllers managed separation assurance. Spacing efficiency was highest when the controllers managed separation assurance but more communications were required.

Combinations of Communications and Aircraft State Many researchers have used combinations of communications and aircraft state. For example, Metzger and Parasuraman (2005) reported a higher percentage of conflicts and self-separations detected in moderate traffic density than in high traffic density. Advance notification times, however, were shorter in high density traffic than in moderate density traffic. Bienert et al. (2014) used delivery accuracy and loss of separation to compare levels of uncertainty in aircraft metering associated with wind and scenario difficulty.

Zingale et al. (2010) used the number of requests to hold aircraft, the number of aircraft in a sector, the number of Push-To-Talk voice transmissions, losses of separation, as well as activation of self-spacing and grouping. They reported no significant differences among Air Traffic Control systems or advanced procedures on requests to hold or redirect aircraft, loss of separation, or self-pacing and grouping. There were significant effects, however, on the number of aircraft managed and number of voice transmissions. Willems and Heiney (2002) reported a significantly lower number of speed changes per aircraft in the low task load scenario than in the high task load scenario. There were no significant differences in the number of altitude changes, heading changes, or handoffs. There was an insufficient number of losses of separation for analysis. The participants were 16 air traffic controllers. Sethumadhavan and Durso (2009) measured map and flight strip recall as well as collisions, violations, and handoff delay in nonradar and radar conditions.

Hannon (2010) used number of departures, taxi time for aircraft, runway occupancy time, and number of transmissions to compare visual and virtual air traffic control towers. There were no significant differences for the number of departures. There were significant effects for taxi times and the number of transmissions. The authors did not state whether the differences in runway occupancy times were significantly different.

Surabattula et al. (2010) used the number and type of interactions (command, information, or request) of controllers (three amateur controllers using Virtual Air Traffic Simulation Network (VATSIM), an online game for controllers and pilots). These were significantly associated with loss of control. Ngo et al. (2012) used handoff delay, en route delay, and errors to compare performance with visual only or visual with auditory, vibrotactile, or audiotactile cues. Errors were loss of aircraft separation; arrival at wrong destination; aircraft at wrong speed, altitude, or heading; or intersection with boundary, another aircraft, or airport. Participants responded more rapidly when the visual cue was augmented with a multisensory cue.

Truitt (2013) compared voice only, 40% data communications, and 75% data communications and reported significant differences in ramp wait time, duration of ground-to-pilot transmissions, number of pilot-to-ground transmissions, duration of pilot-to-ground transmission, and mean time to issue D-Taxi but not in taxi-time out, taxi-time in, number of taxi-out delays, duration of taxi-out delays, or number of ground-to-air radio transmissions.

Data requirements – Rantanen (2004) states that the OTS requires a standardized checklist and extensive training to attain inter-rater reliability.

Thresholds – Not available.

Sources

Askelson, M.A., Dreschel, P., Nordlie, J., Theisen, C.J., Carlson, C., Woods, T., Forsyth, R., and Heitman, R. MQ-9 unmanned aircraft responsiveness to air traffic controller commanded maneuvers: Implications for integration into the National Airspace System. *Air Traffic Control Quarterly* 21(1): 79–92, 2013.

Bienert, N., Mercer, J., Homola, J.R., Morey, S.E., and Prevot, T. Influence of uncertainties and traffic scenario difficulties in a human-in-the-loop simulation. *Air Traffic Control Quarterly* 22(2): 179–193, 2014.

Hadley, G.A., Guttman, J.A., and Stringer, P.G. *Air Traffic Control Specialist Performance Measurement Database (DOT/FAA/CT-TN99/71).* Atlantic City International Airport, NJ: William J. Hughes Technical Center, June 1999.

Hah, S., Willems, B., and Schultz, K. The evaluation of Data Communication for the Future Air Traffic Control System (NextGen). Proceedings of the Human Factors and Ergonomics Society 54th Annual Meeting, 99–103, 2010.

Hannon, D.J. Air traffic controller performance in a human-in-the-loop virtual tower simulation. Proceedings of the Human Factors and Ergonomics Society 54th Annual Meeting, 94–98, 2010.

Khadilkar, H., and Balakrishnan, H. Metrics to characterize airport operational performance using surface surveillance data. *Air Traffic Quarterly* 21(2): 183–206, 2013.

Metzger, U., and Parasuraman, R. Automation in future air traffic management: Effects of decision aid reliability on controller performance and mental workload. *Human Factors* 47(1): 35–49, 2005.

Ngo, M.K., Pierce, R.S., and Spence, C. Using multisensory cues to facilitate air traffic management. *Human Factors* 54(6): 1093–1103, 2012.

Rantanen, E. *Development and Validation of Objective Performance and Workload Measures in Air Traffic Control (AHFD-04-19/FAA-047-7).* Savoy, IL: Aviation Human Factors Division Institute of Aviation University of Illinois, September 2004.

Rovira, E., and Parasuraman, R. Transitioning to future air traffic management: Effects of imperfect automation on controller attention and performance. *Human Factors* 52(3): 411–425, 2010.

Sanchez, J., and Smith, E.C. Tower controllers' response behavior to runway safety alerts. Proceedings of the Human Factors and Ergonomics Society 54th Annual Meeting, 50–54, 2010.

Sethumadhavan, A., and Durso, F.T. Selection in Air Traffic Control: Is nonradar training a predictor of radar performance? *Human Factors* 51(1): 21–34, 2009.

Strybel, T.Z., Vu, K.L., Chiappe, D.L., Morgan, C.A., Morales, G., and Battiste, V. Effects of NextGen concepts of operation for separation assurance and interval management on Air Traffic Controller situation awareness, workload, and performance. *International Journal of Aviation Psychology* 26(1–2), 1–14, 2016.

Surabattula, D., Kaplan, M., and Landry, S.J. Controller intervention to mitigate potential air traffic conflicts. Proceedings of the Human Factors and Ergonomics Society 54th Annual Meeting, 40–44, 2010.

Truitt, T.R. An empirical study of digital taxi clearances for departure aircraft. *Air Traffic Control Quarterly* 21(2): 125–151, 2013.

Vu, K.L., Chiappe, D., Morales, G., Strybel, T.Z., Battiste, V., Shively, J., and Buker, T.J. Impact of UAS pilot communication and execution latencies on air traffic controllers's acceptance of UAS operations. *Air Traffic Control Quarterly* 22(1): 49–80, 2014.
Willems, B., and Heiney, M. *Decision Support Automation Research in the En Route Air Traffic Control Environment (DOT/FAA/CT-TN01/10).* Atlantic City International Airport, NJ: Federal Aviation Administration William J. Hughes Technical Center, January 2002.
Zingale, C.M., Willems, B., and Ross, J.M. Air Traffic Controller workstation enhancements for managing high traffic levels and delegated aircraft procedures. Proceedings of the Human Factors and Ergonomics Society 54th Annual Meeting, 11–15, 2010.

2.4.3 Boyett and Conn's White-Collar Performance Measures

General description – Boyett and Conn (1988) developed lists of performance measures with which to evaluate personnel in white-collar, professional, knowledge-worker organizations. These lists are categorized by the function being performed, i.e., engineering, production planning, purchasing, and management information systems. The lists are provided in Table 2.4.

Strengths and limitation – These measures are among the few developed for white-collar tasks and were developed in accordance with the following guidelines: (1) "involve white-collar employees in developing their own measures" (Boyett and Conn, 1988, p. 210), (2) "measure results, not activities" (p. 210), (3) "use group or team-based measures" (p. 211), and (4) "use a family of indicators" (p. 211).

Data requirements – Not stated.

Thresholds – Not available.

Source

Boyett, J.H., and Conn, H.P. Developing white-collar performance measures. *National Productivity Review.* Summer: 209–218, 1988.

2.4.4 Charlton's Measures of Human Performance in Space Control Systems

General description – Charlton's (1992) measures to predict human performance in space control systems are divided into three phases (pre-pass, contact execution, and contact termination) and three crew positions (ground controller, mission controller, and planner analyst). The measures by phase and crew position are: (1) pre-pass phase, ground-controller time to complete readiness tests and errors during configuration; (2) contact-execution

TABLE 2.4

White-Collar Measures in Various Functions

Engineering
- Percent new or in-place equipment/tooling performing as designed
- Percent machines/tooling capable of performing within established specifications
- Percent operations with current detailed process/method sheets
- Percent work run on specified tooling
- Number of bills of material errors per employee
- Percent engineering change orders per drawing issued
- Percent material specification changes per specifications issued
- Percent engineering change requests to drawings issued based on design or material changes due to drawing/specification errors
- Percent documents (drawings, specifications, process sheets, etc.) issued on time

Production Planning or Scheduling
- Percent deviation actual/planned schedule
- Percent on-time shipments
- Percent utilization manufacturing facilities
- Percent overtime attributed to production scheduling
- Percent earned on assets employed
- Number, pounds, or dollars delayed orders
- Percent back orders
- Percent on-time submission master production plan
- Hours of time lost waiting on materials
- Number of days receipt of work orders prior to scheduled work
- Percent turnover of parts and material (annualized)

Purchasing
- Dollar purchases made
- Percent purchases handled by purchasing department
- Dollar purchases by major type
- Percent purchases/dollar sales volume
- Percent "rush" purchases
- Percent orders exception to lowest bid
- Percent orders shipped "most economical"
- Percent orders shipped "most expeditious"
- Percent orders transportation allowance verified
- Percent orders price variance from original requisition

Purchasing (continued)
- Percent orders "cash discount" or "early payment discount"
- Percent major vendors–annual price comparison completed
- Percent purchases–corporate guidelines met
- Elapsed time–purchase to deliver
- Percent purchases under long-term or "master contract"
- Dollar adjustment obtained/dollar value "defective" or "reject"
- Purchasing costs/purchase dollars
- Purchasing costs/number of purchases
- Dollar value rejects/dollar purchases
- Percent shortages

Management Information Systems
- Number of data entry/programming errors per employee
- Percent reports issued on time
- Data processing costs as percent of sales
- Number of reruns
- Total data processing cost per transaction
- Percent target dates met
- Average response time to problem reports
- Number of data entry errors by type
- Percent off-peak jobs completed by 8:00 am
- Percent end user available (prime) on-line
- Percent on-line 3-second response time
- Percent print turnaround in 1 hour or less
- Percent prime shift precision available
- Percent uninterrupted power supply available
- Percent security equipment available
- Score on user satisfaction survey
- Percent applications on time
- Percent applications on budget
- Correction costs of programming errors
- Number of programming errors
- Percent time on maintenance
- Percent time on development
- Percent budget maintenance
- Percent budget development

From Boyett and Conn (1988, p. 214).

phase, mission-controller track termination time and commanding time; (3) contact-execution phase, planner-analyst communication duration; (4) contact-termination phase, planner-analyst communication duration; and (5) contact-termination phase, ground-controller deconfiguration time, time to return resources, and time to log off system.

Strengths and limitations – The measures were evaluated in a series of three experiments using both civilian and Air Force satellite crews.

Data requirements – Questionnaires are used as well as computer-based scoring sheets.

Thresholds – Not stated.

Source

Charlton, S.G. Establishing human factors criteria for space control systems. *Human Factors* 34: 485–501, 1992.

2.4.5 Driving Parameters

General description – Driving parameters include measures of driver behavior (e.g., average brake RT, brake pedal errors, control light response time, number of brake responses, perception-response time, speed, and steering wheel reversals) as well as measures of total system performance (e.g., time to complete a driving task, tracking error) and observational measures (e.g., vehicles passing, use of traffic lights). Note, Green (2012) urged the use of standard definitions for performance measures and their associated statistics. Also note, the Society of Automotive Engineers (SAE) published operational definitions of driving performance measures and statistics (issued June 30, 2015, standard J2944). In a follow-on study, Liu and Green (2017) compared six measures of mean times to respond to forward collision warnings versus no warnings. Significant differences between the two conditions only occurred when time was measured from the start of the warning to the end of the response.

Strengths and limitations – Most researchers use several driving parameters in a single study. For example, Popp and Faerber (1993) used speed, lateral distance, longitudinal distance to the leading car, acceleration in all axes, steering angle, heading angle, and frequency of head movements of participants driving a straight road in a moving-base simulator to evaluate four feedback messages that a voice command was received. There were no significant differences among these dependent variables as a function of type of feedback message.

2.4.5.1 Average Brake RT

Average brake RT has been used to determine response to events in the traffic environment, evaluate design of brake controls and warnings, and investigate the effect of driver's attributes.

Response to Traffic Events In an early study, Johansson and Rumar (1971) measured the brake RT of 321 drivers in an anticipated event on the road. Five of these were exposed to the same condition two more times. The median brake RT was 0.9 seconds with 25% of the drivers having brake RTs longer than 1.2 seconds. Sivak et al. (1980) reported higher percents of responding to brake lights on both single (54.8%) and dual (53.2%) high-mounted brake lights than conventional brake lights (31.4%). However, there was no difference in mean RT (1.39, 1.30, and 1.38 seconds, respectively). Luoma et al. (1997) reported that yellow turn signals had shorter brake RTs than red turn signals. In a similar study, Sivak et al. (1994) used RT to compare neon, Light Emitting Diode (LED), and fast incandescent brake lamps. There were significant differences among the lamps which were lit in the participant's peripheral vision while the participant was performing a tracking task.

Schrauf et al. (2011) also reported significant increases in brake reaction time while performing a secondary task. Their participants were 20 drivers following a lead vehicle on a closed track. The lead vehicle initiated braking maneuvers at intervals between 42.5 seconds and 57.5 seconds. The secondary task was detecting specific words in an audio book presentation.

In the lead car study, Brookhuis et al. (1994) used RT to speed variations in response to the speed of a leading car as a measure of driver attention. Schweitzer et al. (1995) measured total braking times of 51 drivers 21 to 30 years of age in response to a lead vehicle. There was no significant difference between 60 and 80 kph speeds. There were, however, significant differences between 6 and 12 m following distance (shorter for 6 m) as well as significant differences among naïve, partial knowledge, and full knowledge of the future actions of the lead vehicle (shortest for full knowledge) conditions.

Engstrom et al. (2010) implemented a lead braking vehicle scenario in a driving simulator. They decomposed brake response time into accelerator pedal release time and accelerator-to-brake pedal movement time. There was a significant decrease in accelerator pedal release time with repetition but no significant effect on accelerator-to-brake movement time.

Gugerty et al. (2014) reported that warnings of a green traffic light changing resulted in significantly milder decelerations than without warnings. Their conclusion was based on two simulator studies.

Design of Brake Controls and Warnings In an unusual study, Richter and Hyman (1974) compared the brake RT times for three types of controllers: foot, hand, and trigger. The RT was shortest for the hand controls. In a similar study, Snyder (1976) measured the time to brake for three accelerator-brake pedal distances. He reported longer times for a 6.35 cm lateral and 5.08 cm vertical separation than for 10.16 or 15.24 cm lateral separation with no vertical separation. Also investigating the effects of brake placement, Morrison et al. (1986) found significant differences in movement time from the accelerator to the brake as a function of brake placement (lower than the accelerator resulted in shorter movement times) and gender (women had longer movement times except when the brake was below the accelerator).

Gray (2010) compared seven types of auditory collision warnings using brake time. There were significant differences among them with means varying between ~0.56 to ~0.90 seconds. In another driving simulator, Mohebbi et al. (2009) reported significantly longer braking times for either auditory or tactile warnings while the driver was in a conversation rather than no conversation.

Neubauer et al. (2012) reported faster braking even using a cell phone with full vehicle automation than non-automated vehicle. Kochhar et al. (2012) reported that 100% of participants using a Reverse Braking Aid (RBA) in a test vehicle stopped in 0.6 seconds when presented a true positive signal and 98% of participants in 0.8 seconds after a false positive signal.

Effects of Driver Attributes

Driver attributes include age, training, and amount of rest.

Age. In a comparison of laboratory, stationary, and on-road driving, Korteling (1990) used the RT of correct responses and error percentages. Older drivers (61 to 73 years old) and brain-injured patients had significantly longer RTs than younger drivers (21 to 43 years old). RTs were significantly longer in on-road driving than in the laboratory. There was a significant effect of Inter-Stimulus Interval (ISI). Specifically, the shortest ISI was associated with the longest RT. Patients made significantly more errors than older or younger drivers.

In another study from the lead author, Korteling (1994) used four measures to assess platoon car-following performance: (1) brake RT, (2) the correlation between the speeds of the two cars, (3) time to obtain the maximum correlation, and (4) the maximum correlation. Brake RT and delay time were significantly longer for the patients than for the older or younger drivers. Brake RT was significantly longest when the driving speed of the lead car was varied and the road was winding. Both correlation measures were significantly lower for the patients and older drivers than for the younger drivers.

Szlyk et al. (1995) reported a significant increase in braking response time as age increased. The data were collected in a simulator. There were no significant effects of age or visual impairment on the variability of brake pressure. In another age study, Lambert et al. (2011) reported significantly slower brake reaction times in older than in younger drivers. The data were collected in a driving simulator.

Training. Damm et al. (2011) compared braking times of novice traditionally trained drivers, novice early-trained drivers, and experienced drivers. There were no significant differences among these three groups.

Fitch et al. (2010) measured braking performance of 64 drivers on a closed-course test track. Braking performance was measured as corrected stopping distance, perception time, movement time, stopping time, average deceleration, time to maximum pedal displacement, and maximum pedal force. There were significant effects on expected braking at a barricade. Specifically, older drivers had shorter stopping distance. Time to maximum brake pedal position was shortest for older males and younger females. Further, for drivers in a Volvo S80, movement time was significantly shorter and maximum brake pedal force significantly larger than for drivers of the Mercedes-Benz R350. To an expected auditory alarm, males had significantly shorter stopping distance as did younger drivers and Mercedes-Benz R350 drivers. Male drivers had significantly shorter perception time as did younger drivers. Volvo S80 drivers had significantly shorter stopping time. Finally, males had significantly larger maximum brake pedal force as did younger drivers and Volvo S80 drivers. Meng et al. (2015) used vibrotactile signals for forward collision warning. They reported shorter RTs when the warning was toward the torso rather than away from the torso.

Gaspar et al. (2012) used overall response time to pedestrian crossing, cars pulling out, cars turning in front, and dogs crossing to evaluate a commercial training package for older drivers. There were no significant differences in these measures between the training package group and a control group.

Amount of Rest. Drory (1985) reported significant differences in average brake RT associated with different types of secondary tasks. It was not affected by the amount of rest drivers received prior to the simulated driving task.

2.4.5.2 Brake Pedal Errors

Rogers and Wierwille (1988) used the type and frequency of pedal actuation errors to evaluate alternative brake pedal designs. They reported 297 errors over 72 hours of testing. Serious errors occurred when the wrong or both pedals were depressed. Catch errors occurred when a pedal interfered with a foot movement. If the interference was minimal, the error was categorized as a scuff. Instructional errors occurred when the participant failed to perform the task as instructed.

Vernoy and Tomerlin (1989) used pedal error, "hitting the accelerator pedal when instructed to depress the brake pedal" (p. 369), to evaluate misperception of the centerline. There were no significant differences in pedal error among eight types of automobile used in the evaluation. There was a significant difference in deviation from the centerline among the eight automobiles, however. Finally, Wu et al. (2014) reported significantly higher pedal rates for older drivers than for younger drivers.

2.4.5.3 Control Light Response Time

Drory (1985) reported significant differences in control light response time associated with various types of secondary tasks. Time was not affected by the amount of rest drivers received prior to the simulated driving task. Although not a control light, Summala (1981) placed a small lamp on roads and measured the RT of drivers for initiating an avoidance maneuver. RTs were about 2.5 seconds with an upper safe limit projected at three seconds. A related measure is the time to take manual control from an autonomous vehicle (Funkhouser and Drews, 2016).

2.4.5.4 Number of Brake Responses

Drory (1985) used the number of brake responses to evaluate types of secondary tasks. The type of task did not affect the number of brake responses nor did the amount of rest drivers received prior to the simulated driving task. Gray (2010) compared seven types of auditory collision warnings using mean number of brake activations. There was a significant effect of warning type (constant intensity, pulsed, ramped, car horn, loom late, loom veridical, and loom early). In the same year, Morgan and Hancock (2010) reported significantly more braking inputs after a navigation system failed than at the beginning of the drive. The data were collected in a driving simulator.

Attwood and Williams (1980) reported a significant increase in brake reversal rate under alcohol than under either cannabis or cannabis and alcohol conditions. The data were collected from eight male drivers on a closed course in an instrumented vehicle.

2.4.5.5 Number of Collisions

Number of collisions is typically used to evaluate collision avoidance systems, road design, and driver training.

Collision Avoidance Systems. Gray (2010) compared seven types of auditory collision warnings. There was a significant difference of warning type (constant intensity, pulsed, ramped, car horn, loom late, loom veridical, and loom early). Drew and Hayes (2010) used the time between oncoming traffic in an intersection and the vehicle's rear bumper exiting the lane during a turn through the intersection. There were no significant effects in the near lanes of icons, mirrors, and vibrotactile seats. For the far lanes, however, there were significantly longer times associated with the seat than with the icons.

Stanton et al. (2011) wanted to use the number of collisions to compare performance of drivers with a Graded Deceleration Display versus the standard brake lights. With the new display, there were only five collisions over the 485 trials. In a test track evaluation of collision avoidance systems (forward collision warning and lane-change merge warning), Fitch et al. (2014)

reported significantly faster maneuvering away from the lateral crash when participants received both warnings than when they received only the forward collision warning.

In a novel study, Xiong and Zhao (2016) reported that taxi drivers using two taxi-hailing apps at the same time had a shorter time to avoid a collision than drivers not using the apps. Beller et al. (2013) compared the time to collision for drivers presented uncertainty estimates of lead vehicle intent with those not given the estimates. The former had a significantly longer minimum time to collision. The data were collected in a simulator. Also in a simulator, Inman et al. (2016) reported a significantly longer time to collision with cooperative adaptive cruise control than without it. Muslim and Itoh (2017) compared the number of side collisions or rear-end collisions that occurred in a driving simulator with a haptic lane change collision avoidance system versus an automatic lane change collision avoidance system. The data were collected from 48 licensed drivers with three types of hazardous lane change scenarios: (1) blind spot hazard, (2) fast approaching follow vehicle, and (3) blind spot with suddenly stopping lead vehicle. Both systems reduced the number of side accidents in scenarios 1 and 3 but increased the number in scenario 2.

Road Design. Fitch and Hankey (2012) examined near crash lane changes in a naturalistic driving study. They reported a significantly shorter time to cross lane markings during right lane changes resulting in near crashes than those right lane changes not near crash. They were also significantly less likely to use the turn signal in the near crash condition. The left lane change comparisons between near crash and not near crash were not significantly different in time to cross lane markings. There were no differences in turn signal use for left lane changes near crash or not.

Wemeke et al. (2014) estimated the timing required by drivers for collision warnings. There was no significant effect of type of road way (urban, rural, highway) on time to collision but there was for timing of the warning. For distance, both main effects and the interaction were significant. Also using time to collision, Lee et al. (2016) reported significantly longer time to collision when there was no reduction in visual field from when there was a reduction.

Driver Training. Damm et al. (2011) compared number of collisions of novice traditionally trained drivers, novice early-trained drivers, and experienced drivers. The authors reported that half of the collisions occurred within the traditionally trained group. In a unique application, Helton et al. (2014) used number of collisions to evaluate performance of operators of Unmanned Ground Vehicles (UGVs). Read and Saleem (2017) reported a significant increase in the number of collisions in a parking task when drivers were trained using either a virtual reality head-mounted display or a flat screen than when trained in the real world.

2.4.5.6 Perception-Response Time

Olson and Sivak (1986) measured perception-response time from the first sighting of an obstacle until the accelerator was released and the driver contacted the brake. Their data were collected in an instrumented vehicle driven on a two-lane rural road.

2.4.5.7 Speed

Speed has been used to evaluate driver characteristics, road configurations, driving conditions, nondriving tasks, vehicle characteristics, and perception.

Driver Characteristics. In a simulator study, Szlyk et al. (1995) reported a significant decrease in speed as age increased. In another age study, Lambert et al. (2010) reported no significant difference in speed between older than in younger drivers. The data were collected in a driving simulator. Reimer et al. (2012) reported significant decrease in speed as secondary task demand increased. Further, older drivers drove slower than younger drivers. The data were collected in an instrumented vehicle on actual roadways.

Damm et al. (2011) compared mean speed of novice traditionally trained drivers, novice early-trained drivers, and experienced drivers. There were no significant differences among these three groups in initial speed. However, there were time effects: an initial decrease then increase at the end of the overtaking scenario, the opposite vehicle crossing scenario, the left crossroads scenario, and parked vehicle scenario. There was a similar but not significant effect in the pedestrian scenario. Krasnova et al. (2016) used mean maximum speed to evaluate the effectiveness of three speeding interventions (pretraining, immediate post-training, one-week follow up) on teen drivers.

In a recent study de Groot et al. (2011) used speed to evaluate the effects of on-target and off-target feedback. The data were collected in a driving simulator. The participants were 60 persons without drivers' licenses. There were significant effects for speed (slower during practice 2 than during practice 1 and 3).

Attwood and Williams (1980) reported significant increases in both mean and median speed when drivers were under the influence of cannabis and a significant decrease when drivers were under the influence of both alcohol and cannabis. The data were collected from eight male drivers on a closed course in an instrumented vehicle.

Road Configurations. Steyvers and de Waard (2000) reported a significant increase in speed on roads with edgelines than in the absence of edgelines. In another edgeline study, Steyvers and de Waard (2000) reported a significant increase in speed on roads with edgelines than in the absence of edgelines. In an on-road study, Nowakowski and Sharafsaleh (2013) reported a significant decrease in speed when a roadside animal crossing sign was illuminated.

In a simulator study, Jeong and Liu (2017) reported significantly higher mean speed when drivers were following a vehicle than when they were in free flow driving. Mean speed was also higher in a moderate curve (800 m radius) than in a sharp curve (100 m radius).

Driving Conditions. Applying a series of speed measures, Shinar and Stiebel (1986) used speed, speed above or below the speed limit, speed reduction (((original speed − speed at site 1)/original speed) × 100) and speed resumption (((speed at site 2 − speed limit)/speed limit) × 100) to evaluate the effectiveness of the presence of police cars in reducing speeding. In an unusual simulator study, Chan et al. (2014) reported significant decreases in speed and rmse lane position when billboards with taboo images were present than when they were not.

Nondriving Tasks. Using both a simulator and an old road vehicle, Reed and Green (1999) reported a significant increase in lateral speed when dialing a telephone while driving on the road or in a simulator. In another simulator study, McGough and Ma (2015) reported significantly smaller speed deviation when a phone was located on the dashboard than in the cupholder, pillar, or shifter.

Jongen et al. (2011) reported that distraction was associated with reduced speed. The data were collected in a driving simulator. In another simulator study, Lee et al. (2012) reported a significant increase in speed variability when drivers searched long playlists (580 songs) rather than short playlists (20 songs). Their data were from 50 drivers in a fixed based, medium fidelity simulator.

Vehicle Characteristics. Wilson and Anderson (1980) compared driving speeds on a closed course and in traffic using radial or cross-ply tires. They reported no significant correlation of mean speed on the track and on the road. There was a significant correlation between speed on the test track and age of the driver, however. There were also significant increases in speed over trials and with the radial versus the cross-ply tire.

Morgan and Hancock (2010) also reported a significant effect of speed in a driving simulator. They reported slower speed before failure of a navigation system than immediately after or at the end of the drive. Similar results were reported for speed differential. Clark and Feng (2015) reported significant differences in minimum speeds at which drivers took control of semi-autonomous vehicles when approaching a construction zone. Speed increased with practice. The data were collected in a driving simulator. In an instrumented vehicle study, Reagan et al. (2013) reported significant reductions in driving faster than the posted speed limit when a speed alerting system was active.

Perception. Svenson (1976) reported that participants overestimated speed of slower segments if faster segments were included. Their participants made their estimates from model trains but the results were generalized to driving. In an on-road study, Salmon et al. (2011) reported that speeding was the most frequent driving error.

2.4.5.8 Steering Wheel Reversals

Steering wheel reversals have been used for decades driving difficulty, driver state, and vehicle characteristics.

Driving Difficulty. Hicks and Wierwille (1979) reported that steering reversals were sensitive to workload (i.e., gusts at the front of a driving simulator). He and McCarley (2011) also reported higher rate in higher cognitively demanding task. Their data were collected in a driving simulator. Reimer et al. (2012) reported wheel reversal rates increased in greater cognitive demand conditions. Further, older drivers had higher reversal rates than younger drivers. The data were collected in an instrumented vehicle on actual roadways.

Driver State. Drory (1985) reported significant differences in steering wheel reversals associated with various types of secondary tasks. It was not affected by the amount of rest drivers received prior to the simulated driving task, however. Similarly, Horne and Baumber (1991) reported no effect of either time of day or alcohol on lateral corrective steering movements. de Groot et al. (2011) used steering wheel reversal to evaluate the effects of on-target and off-target feedback. They reported that steering wheel reversal rate decreased over practice sessions.

Vehicle Characteristics. Frank et al. (1988) used the number of large steering reversals (greater than 5 degrees), number of small steering reversals (less than 5 degrees), and yaw standard deviation ("angle in the horizontal plane between the simulated vehicle longitudinal axis and the instantaneous roadway tangent," p. 206) to evaluate the effects of motion system transport delay and visual system transport delay. All three measures of driver performance were significantly related to transport delay.

In another driving simulator study, Mulder et al. (2012) compared manual versus haptic shared control. They reported a 16% decrease in steering wheel reversal rate as well as a 15% decrease in standard deviation of steering wheel angle in the haptic shared control condition than in the manual control condition. In a driving simulator, Samost et al. (2015) reported significantly more major steering wheel reversals when drivers used a smartphone in the visual-manual calling method than in the auditory-vocal calling method using either a Smartphone or Smartwatch.

2.4.5.9 Time

Time has been extensively used to evaluate driver characteristics, driving task, road design, response to traffic conditions, and vehicle design.

Driver Characteristics. Stahl et al. (2014) reported no significant correlation between drivers' self-rating as safe drivers and time to collisions.

Driving Task. Finnegan and Green (1990) reviewed five studies in which time to lane crossing (TLC) was measured. For these studies, the authors concluded that 6.6 seconds are required for the visual search with a single lane change and 1.5 seconds to complete the lane change. With yet another measure, Reed and Green (1999) reported that lane keeping was less precise in a simulator than on the road. Lane keeping was also less precise when dialing a telephone during driving in either the simulator or on the road.

Road Design. Godthelp (1986) reported, based on field study data, that TLC described anticipatory steering action during curve driving. In a later study, Godthelp (1988) used TLC to assess the effects of speed on driving strategies from error neglecting to error correcting at lane edges.

Traffic Conditions. Sidaway et al. (1996) asked participants to estimate time to collision after viewing videotapes of accidents. Participants consistently underestimated the time. However, as velocity increased the time estimate was more accurate.

Vehicle Design. Gawron et al. (1986) used the time to complete a double-lane change as well as the number of pylons struck to evaluate side impact padding thickness (0, 7.5, 10 cm), direction of the lane change (left/right, right/left), and replication (1 to 12). Participants took longer to perform the left/right than the right/left lane change. Time to complete the lane change increased as padding thickness increased (0 cm = 3.175 s, 7.5 cm = 3.222 s, and 10 cm = 3.224 s). There were no significant effects on the number of pylons struck.

Roge (1996) used TLC as a measuring of steering quality. He reported significantly better steering control among participants who were frame independent. Others have used TLC to enhance preview-predictor models of human driving performance. In these models, TLC equals the time for the vehicle to reach either edge of the driving lane. It is calculated from lateral lane position, the heading angle, vehicle speed, and commanded steering angle (Godthelp et al., 1984). Godthelp et al. (1984) evaluated TLC in an instrumented car driven on an unused straight, four-lane highway by six male drivers at six different speeds (20, 40, 60, 80, 100, and 120 km/hr) with and without a visor. Based on the results, the authors argued that TLC was a good measure of open-loop driving performance.

Lamble et al. (1999) used a time to collision to evaluate locations of in-vehicle displays and controls. Also using time to collision, van Winsum and Heino (1996) reported that time-to-collision information was used to initiate and control braking. Their data were also collected in a driving simulator. Their participants were 54 male drivers with an average age of 29 years.

Yager et al. (2012) reported significantly longer time to respond to a peripheral light mounted on a car driving in a closed course while texting and driving than driving without texting. In an extensive study, Gold et al. (2013) compared length of time to respond to a takeover request presented auditorily and visually in a highly automated driving simulator or highly automated automobile on a test track. The authors concluded that shorter times to take over generally resulted in faster decisions.

Eriksson and Stanton (2017) provided an excellent review of research measuring transitions to and from manual control. These authors also collected data from 26 drivers in a fixed-based driving simulator located at the University of Southampton. Drivers experienced three conditions: (1) manual, (2) highly automated driving (HAD), and (3) HAD with a secondary task (reading a magazine). Without a secondary task, drivers took up to 25.750 seconds to transition from automated to manual and 23.884 seconds from manual to automated. With a secondary task, maximum times were shorter: 20.997 (automated to manual) and 23.221 seconds (manual to automated).

As highly automated vehicles become more available, time to take over from the automation is being used to evaluate the human machine interface. Time variables include time to react to alert, time to regain control, time to release control (this includes both time to activate the automation and the time to release control of steering) (Blanco et al., 2015).

2.4.5.10 Tracking Error

Tracking error has been used to evaluate response to driver characteristics, driving conditions, and road design.

Driver Characteristics. Korteling (1994) reported no significant differences in the standard deviation of lateral position between young (21–34) and old (65–74 years) drivers in a driving simulator. However, older drivers had significantly larger longitudinal standard deviation in a car-following task. There was also a decrement associated with fatigue in steering performance but not in car-following performance. In another age study, Lambert et al. (2010) reported a trend toward larger following distances in older than in younger drivers. The data were collected in a driving simulator. Gaspar et al. (2012) used mean following distance, headway safety margin, and tailway safety margin to evaluate a commercial training package for older drivers. There were no significant differences in these measures between the training package group and a control group.

Hollopeter et al. (2012) reported that novice male drivers had significantly greater variability in lane keeping than novice female drivers or experienced male drivers. The data were collected in a driving simulator. In another measure of deviation, Szlyk et al. (1995) reported a significant increase in the number of lane crossings as age increased. The data were collected in a simulator. In a related study, Summala et al. (1996) reported worse lane-keeping performance for novice drivers than for experienced drivers when the display was near the periphery rather than in the middle console of a driving simulation.

Damm et al. (2011) compared lane position of novice traditionally trained drivers, novice early-trained drivers, and experienced drivers. The experienced group was significantly more to the left in the lane than the traditional group in an overtaking scenario and the traditional group in the parked vehicle scenario. The experienced group was significantly more right than the early group in the parked vehicle scenario as well. There was no significant group effect in the pedestrian scenario, the opposite vehicle crossing scenario, or the left crossroads scenario.

Using a driving simulator, Drory (1985) reported significant differences in tracking error associated with various types of secondary tasks. It was not affected by the amount of rest drivers received prior to the simulated driving task. Atchley and Chan (2011) reported enhanced lane keeping performance when performing a concurrent verbal task. Their data were collected in a driving simulator. Hosking and Young (2009) reported that use of a cell phone to retrieve and send text messages increased the variability of both the following distance to lead vehicles and variability in lane position. Libby and Chaparro (2009) also reported an increase in variability in lane position but with text messaging during driving versus talking on a cell phone. In a similar study, Lee et al. (2012) reported a significant increase in the standard deviation of lane position when drivers searched long playlists (580 songs) rather than short playlists (20 songs). Their data were from 50 drivers in a fixed based, medium fidelity simulator.

Xiong et al. (2012) used minimum time headway to compare conservative, moderate, and risky drivers.

Lane keeping performance was also used by de Groot et al. (2011) to evaluate the effects of on-target and off-target feedback. Their measure of lane deviation was percent time that the vehicle was within 0.5 meters of the center of the road. The data were collected in a driving simulator. The participants were 60 persons without drivers' licenses. Another measure of lane deviation was the number of road departures. They reported significant improvement in percent time on target for the group without augmented feedback than for the group over time. There were also significant differences in rmse off target versus nonaugmented feedback and off target versus on target. There were no significant effects on number of departures.

Using other related measures, Heimstra et al. (1980) used time on target and frequency off target to assess the effects of cigarette deprivation on smokers' tracking performance. There were significant effects on both measures. Horne and Baumber (1991) reported significant effects of alcohol on average following distance in a driving simulator.

Response to Driving Conditions. In an early study, Hicks and Wierwille (1979) reported that yaw deviation and lateral deviation were sensitive to workload (i.e., gusts at the front of a driving simulator). Also in a simulator, Medeiros-Ward et al. (2010) reported that drivers made significantly more lane crossings in low (just drive) and medium (drive and perform digit classification task) workload conditions than in their high (drive and backward counting by threes) workload condition. In addition, there was significantly less lateral distance traveled as workload increased. Finally, lateral lane position variability was greatest in the low workload condition.

In another driving simulator, He et al. (2011) reported that drivers drove farther to the right and had greater variability in lane deviation in heavy wind than in no wind conditions. There was also significantly less variability in speed when participants reported that they were "mind wandering" than when attentive. There were no significant effects of wind or mind wandering on mean velocity, headway distance to lead car, and time to contact lead car. He and McCarley (2011) reported significantly less lane variability (standard deviation of lane position) when driving a simulator under high cognitive load.

In a naturalistic study of heavy truck drivers' following behavior, Bao et al. (2012) reported significantly longer time headways (0.28 s) in dense traffic when an in-vehicle crash warning system was installed. There was a similar increase with the wipers on (0.20 s).

Imbeau et al. (1989) reported that if drivers failed to respond to a display reading task, the variance of lane deviation decreased. The data were collected in a driving simulator. In another simulator study, Dijsterhuis et al. (2012) compared lane keeping performance when lateral position feedback was provided on the simulator HUD all the time (nonadaptive), when performance indicated the need (adaptive), and not at all (no support). The 31 experienced drivers in the study tended to drive nearer the center of the road in the adaptive than the nonadaptive case ($p = 0.051$), had significantly smaller standard deviation in lane position in the adaptive than the nonadaptive mode, and spent 3% less time driving outside the driving lane edges in the adaptive than in the nonadaptive mode. Skottle et al. (2014) in a simulator study reported significant increase in the standard deviation of lateral position during manual driving after driving in a highly automated vehicle. In a comparison of different levels of automation (lane keeping alone versus lane keeping with adaptive cruise control), Shen and Neyens (2014) reported significantly larger maximum lane deviations at higher levels of automation.

Godthelp and Kappler (1988) reported larger standard deviations in lateral position when drivers were wearing safety goggles than when they were not. This measure also increased as speed increased. In an unusual application, evaluating underground mining vehicles, Cloete et al. (2012) reported significant differences in lateral deviation associated with differences in control order of joysticks used in these vehicles. Petermeijer et al. (2014) reported significant differences in lateral error associated with different levels of haptic steering support. The participants were 32 licensed drivers in a medium fidelity driving simulator.

Crandall and Chaparro (2012) used mean vehicle deviation from the optimal path to compare the effect of texting (no texting, texting with physical keyboard, texting with touch screen). There was significantly larger deviation in the touch keyboard condition than in the physical keyboard and this was greater than in the no texting condition. Yager et al. (2012) reported significantly larger standard deviation of lane position in a car driving in a closed course while writing and reading texts and driving than driving without writing and reading texts. Using the standard deviation of lane position, Samost et al. (2015) reported significantly greater standard deviations when drivers used a smartphone in visual-manual calling method than in the auditory-vocal calling method using either Smartphone or Smartwatch.

Stanton et al. (2011) used minimum following distance to compare performance of drivers with a Graded Deceleration Display versus the standard brake lights. There was significantly safer following distance with the standard brake lights than with the Graded Deceleration Lights.

Road Design. In an early study, Soliday (1975) measured lane position maintenance of 12 male drivers on the road. Drivers tended to stay in the center of the lane and oscillated around the center of two-lane roads more than four-lane roads. van Winsum (1996) reported that steering wheel angle increased as curve radii decreased. Steering error increased as steering wheel angle increased. The data were collected in a driving simulator.

2.4.5.11 Observational Measures

Wooller (1972) observed four drivers negotiate an 11.72 mile route to determine the following measures: (1) travel time, (2) vehicles passing participant, (3) vehicles passed by participant, (4) lateral position changes, (5) use of traffic lights, and (6) number of vehicles. Patterns varied across participants but were consistent within a participant. Aksan et al. (2013) compared average miles per hour, lateral acceleration, longitudinal acceleration, and steering of middle-aged and older drivers in an instrumented vehicle. Older drivers were significantly slower and applied less lateral and longitudinal acceleration. There was no significant effect of age on steering, however. Older drivers also had larger proportions of lane change, lane observance, control of speed, turns, and turn signal errors than middle-aged drivers.

Data requirements – An instrumented simulator or vehicle is required.

Thresholds – Total time varied between 0.1 and 1.8 seconds (Olson and Sivak, 1986).

Sources

Aksan, N., Dawson, J.D., Emerson, J.L., Yu, L., Uc, E.Y., Anderson, S.W., and Rizzo, M. Naturalistic distraction and driving safety in older drivers. *Human Factors* 55(4): 841–853, 2013.

Atchley, P., and Chan, M. Potential benefits and costs of concurrent task engagement to maintain vigilance: A driving simulator investigation. *Human Factors* 53(1): 3–12, 2011.

Attwood, D.A., and Williams, R.D. Braking performance of drivers under the influence of alcohol and cannabis. Proceedings of the Human Factors Society 24th Annual Meeting, 134–138, 1980.

Bao, S., LeBlanc, D.J., Sayer, J.R., and Flannagan, C. Heavy-truck drivers' following behavior with intervention of an integrated, in-vehicle crash warning system: A field evaluation. *Human Factors* 54(5): 687–697, 2012.

Beller, J., Heesen, M., and Vollrath, M. Improving the driver-automation interaction: An approach using automation uncertainty. *Human Factors* 55(6): 1130–1141, 2013.

Blanco, M., Atwood, J., Vasquez, H.M., Trimble, T.E., Fitchett, V.L., Radlbeck, J., Fitch, G.M., Russell, S.W., Green, C.A., Cullinane, B., and Morgan, J.F. *Human Factors Evaluation of Level 2 and Level 3 Automated Driving Concepts (DOT HS 812 182).* Washington, DC: National Highway Traffic Safety Administration, August 2015.

Brookhuis, K., de Waard, D., and Mulder, B. Measuring driving performance by car-following in traffic. *Ergonomics* 37(3): 427–434, 1994.

Chan, M., Madan, C.R., and Singhal, A. The effects of taboo-related distraction on driving performance. Proceedings of the Human Factors and Ergonomics Society 58th Annual Meeting, 1366–1370, 2014.

Clark, H., and Feng, J. Semi-autonomous vehicles: Examining driver performance during the take-over. Proceedings of the Human Factors and Ergonomics Society 59th Annual Meeting, 781–785, 2015.

Cloete, S., Zupanc, C., Burgess-Limerick, R., and Wallis, G. Steering performance and dynamic complexity in a simulated underground mining vehicle. Proceedings of the Human Factors and Ergonomics Society 56th Annual Meeting, 1341–1345, 2012.

Crandall, J.M., and Chaparro, A. Driver distraction: Effects of text entry methods on driving performance. Proceedings of the Human Factors and Ergonomics Society 56th Annual Meeting, 1693–1697, 2012.

Damm, L., Nachtergaele, C., Meskali, M., and Berthelon, C. The evaluation of traditional and early driver training with simulated accident scenarios. *Human Factors* 53(4): 323–337, 2011.

de Groot, S., de Winter, J.C.F., Garcia, J.M.L., Mulder, M., and Wieringa, P.A. The effect of concurrent bandwidth feedback on learning the lane-keeping task in a driving simulator. *Human Factors* 53(1): 50–62, 2011.

Dijsterhuis, C., Stuiver, A., Mulder, B., Brookhuis, K.A., and de Waard, D. An adaptive driver support system: User experiences and driving performance in a simulator. *Human Factors* 54(5): 772–785, 2012.

Drew, D.A., and Hayes, C.C. In-vehicle decision support to reduce crashes at rural thru-stop intersections. Proceedings of the Human Factors and Ergonomics Society 54th Annual Meeting, 2028–2032, 2010.

Drory, A. Effects of rest and secondary task on simulated truck-driving task performance. *Human Factors* 27(2): 201–207, 1985.

Engstrom, J., Aust, M.L., and Vistrom, M. Effects of working memory and repeated scenario exposure on emergency braking performance. *Human Factors* 52(5): 551–559, 2010.

Eriksson, A., and Stanton, N.A. Takeover time in highly automated vehicles: Noncritical transitions to and from manual control. *Human Factors* 59(1): 1–17, 2017.

Finnegan, P., and Green, P. *The Time to Change Lanes: A Literature Review (UMTRI-90–34).* Ann Arbor, Michigan: The University of Michigan Transportation Research Institute, September 1990.

Fitch, G.M., Blanco, M., Morgan, J.F., and Wharton, A.E. Driver braking performance to surprise and expected events. Proceedings of the Human Factors and Ergonomics Society 54th Annual Meeting, 2076–2080, 2010.
Fitch, G.M., Bowman, D.S., and Llaneras, R.E. Distracted driver performance to multiple alerts in a multiple-conflict scenario. *Human Factors* 56(8): 1497–1505, 2014.
Fitch, G.M., and Hankey, J.M. Investigating improper lane changes: Driver performance contributing to lane change near crashes. Proceedings of the Human Factors and Ergonomics Society 56th Annual Meeting, 2231–2235, 2012.
Frank, L.H., Casali, J.G., and Wierwille, W.W. Effects of visual display and motion system delays on operator performance and uneasiness in a driving simulator. *Human Factors* 30(2): 201–217, 1988.
Funkhouser, K., and Drews, F. Reaction times when switching from autonomous to manual driving control: A pilot investigation. Proceedings of the Human Factors and Ergonomics Society 60th Annual Meeting, 1847–1851, 2016.
Gaspar, J.G., Neider, M.B., Simons, D.J., McCarley, J.S., and Kramer, A.F. Examining the efficacy of training interventions in improving older driver performance. Proceedings of the Human Factors and Ergonomics Society 56th Annual Meeting, 144–148, 2012.
Gawron, V.J., Baum, A.S., and Perel, M. Effects of side-impact padding on behavior performance. *Human Factors* 28(6), 661–671; 1986.
Godthelp, H. Vehicle control during curve driving. *Human Factors* 28(2): 211–221, 1986.
Godthelp, H. The limits of path error neglecting in straight lane driving. *Ergonomics* 31(4): 609–619, 1988.
Godthelp, H., and Kappler, W.D. Effects of vehicle handling characteristics on driving strategy. *Human Factors* 30(2): 219–229, 1988.
Godthelp, H., Milgram, P., and Blaauw, G.J. The development of a time-related measure to describe driving strategy. *Human Factors* 26(3): 257–268, 1984.
Gold, G., Dambock, D., Lorenz, L., and Bengler, K. "Take over!" how long does it take to get the driver back into the loop? Proceedings of the Human Factors and Ergonomics Society 57th Annual Meeting, 1938–1942, 2013.
Gray, R. Looming auditory collision warnings for driving. *Human Factors* 53(1): 63–74, 2010.
Green, P. Using standards to improve the replicability and applicability of driver interface research. Proceedings of the 4th International Conference on Automotive User Interfaces and Interactive Vehicular Applications (AutomotiveUI '12), 15–22, 2012.
Gugerty, L., McIntyre, S.E., Link, D., Zimmerman, K., Tolani, D., Huang, P., and Pokorny, R.A. Effects of intelligent advanced warnings on drivers negotiating the Dilemma Zone. *Human Factors* 56(6): 1021–1035, 2014.
He, J., Becic, E., Lee, Y., and McCarley, J.S. Mind wondering behind the wheel: Performance and occulometer correlates. *Human Factors* 53(1): 13–21, 2011.
He, J., and McCarley, J.S. Effects of cognitive distraction on lane-keeping performance loss or improvement? Proceedings of the Human Factors and Ergonomics Society 55th Annual Meeting, 1894–1898, 2011.
Heimstra, N.W., Fallesen, J.J., Kinsley, S.A., and Warner, N.W. The effects of deprivation of cigarette smoking on psychomotor performance. *Ergonomics* 23(11): 1047–1055, 1980.
Helton, W.S., Head, J., and Blaschke, B.A. Cornering law: The difficulty of negotiating corners with an unmanned ground vehicle. *Human Factors* 56(2): 392–402, 2014.

Hicks, T.G., and Wierwille, W.W. Comparison of five mental workload assessment procedures in a moving-base driving simulator. *Human Factors* 21(2): 129–143, 1979.

Hollopeter, N., Brown, T., and Thomas, G. Differences in novice and experienced driver response to lane departure warnings that provide active intervention. Proceedings of the Human Factors and Ergonomics Society 56th Annual Meeting, 2216–2220, 2012.

Horne, J.A., and Baumber, C.J. Time-of-day effects of alcohol intake on simulated driving performance in women. *Ergonomics* 34(11): 1377–1383, 1991.

Hosking, S.G., and Young, K.L. The effects of text messaging on young drivers. *Human Factors* 51(4): 582–592, 2009.

Imbeau, D., Wierwille, W.W., Wolf, L.D., and Chun, G.A. Effects of instrument panel luminance and chromaticity on reading performance and preference in simulated driving. *Human Factors* 31(2): 147–160, 1989.

Inman, V.W., Jackson, S., and Philips, B.H. Driver performance in a cooperative adaptive cruise control string. Proceedings of the Human Factors and Ergonomics Society 60th Annual Meeting, 1183–1187, 2016.

Jeong, H., and Liu, Y. Horizontal curve driving performance and safety affected by road geometry and lead vehicle. Proceedings of the Human Factors and Ergonomics Society Annual Meeting, 1629–1633, 2017.

Johansson, G., and Rumar, K. Driver's brake reaction times. *Human Factors* 13(1): 23–27, 1971.

Jongen, E.M.M., Brijs, K., Mollu, K., Brijs, T., and Wets, G. 70 km/h speed limits on former 90 km/h roads: Effects of sign repetition and distraction on speed. *Human Factors* 53(6): 771–785, 2011.

Kochhar, D.S., Talamonti, W.J., and Tijerina, L. Driver response to unexpected automatic/haptic warning while backing. Proceedings of the Human Factors and Ergonomics Society 56th Annual Meeting, 2211–2215, 2012.

Korteling, J.E. Perception-response speed and driving capabilities of brain-damaged and older drivers. *Human Factors* 32(1): 95–108; 1990.

Korteling, J.E. Effects of aging, skill modification, and demand alternation on multiple task performance. *Human Factors* 36(1): 27–43, 1994.

Krasnova, O., Molesworth, B., and Williamson, A. Understanding the effect of feedback on young drivers' speeding behavior. Proceedings of the Human Factors and Ergonomics Society 60th Annual Meeting, 1979–1983, 2016.

Lambert, A.E., Watson, J.M., Cooper, J.M., and Strayer, D.L. The roles of working memory capacity, visual attention and age in driving performance. Proceedings of the Human Factors and Ergonomics Society 54th Annual Meeting, 170–174, 2010.

Lamble, D., Laakso, M., and Summala, H. Detection thresholds in car following situations and peripheral vision for positioning of visually demanding in-car displays. *Ergonomics* 42(6): 807–815, 1999.

Lee, J., Itoh, M., and Inagaki, T. Effectiveness of driver compensation to avoid vehicle collision under visual field construction. Proceedings of the Human Factors and Ergonomics Society 60th Annual Meeting, 1904–1908, 2016.

Lee, J.D., Roberts, S.C., Hoffman, J.D., and Angell, L.S. Scrolling and driving: How an MP3 player and its aftermarket controller affect driving performance and visual behavior. *Human Factors* 54(2): 250–263, 2012.

Libby, D., and Chaparro, A. Text messaging versus talking on a cell phone: A comparison of their effects on driving performance. Proceedings of the Human Factors and Ergonomics Society 53rd Annual Meeting, 1353–1357, 2009.
Liu, K., and Green, P. The conclusion of a driving study about warnings depends upon how response time is measured. Proceedings of the Human Factors and Ergonomics Society Annual Meeting, 1876–1880, 2017.
Luoma, J., Flannagan, M.J., Sivak, M., Aoki, M., and Traube, E.C. Effects of turn-signal colour on reaction time to brake signals. *Ergonomics* 40(1): 62–68, 1997.
McGough, B., and Ma, W. Assessment of in-vehicle cellphone locations in influencing driving performance and distraction. Proceedings of the Human Factors and Ergonomics Society 59th Annual Meeting, 1588–1592, 2015.
Medeiros-Ward, N., Seegmiller, J., Cooper, J., and Strayer, D. Dissociating eye movements and workload on lateral lane position variability. Proceedings of the Human Factors and Ergonomics Society 54th Annual Meeting, 2067–2070, 2010.
Meng, F., Gray, R., Ho, C., Ahtamad, M., and Spence, C. Dynamic vibrotactile signals for forward collision avoidance warning systems. *Human Factors* 57(2): 329–346, 2015.
Mohebbi, R., Gray, R., and Tan, H.Z. Driver reaction time to tactile and auditory rear-end collision warnings while talking on a cell phone. *Human Factors* 51(1): 102–110, 2009.
Morgan, J.F., and Hancock, P.A. The effect of prior task loading on mental workload: An example of hysteresis in driving. *Human Factors* 53(1): 75–86, 2010.
Morrison, R.W., Swope, J.G., and Malcomb, C.G. Movement time and brake pedal placement. *Human Factors* 28(2): 241–246, 1986.
Mulder, M., Abbink, D.A., and Boer, E.R. Sharing control with haptics: Seamless driver support from manual to automatic control. *Human Factors* 54(5): 786–798, 2012.
Muslim, H., and Itoh, M. Human factor issues associated with lane change collision avoidance systems: Effects of authority, control, and ability on drivers' performance and situation awareness. Proceedings of the Human Factors and Ergonomics Society Annual Meeting, 1634–1638, 2017.
Neubauer, C., Matthews, G., and Saxby, D. The effects of cell phone use and automation on driver performance and subjective state in simulated driving. Proceedings of the Human Factors and Ergonomics Society 56th Annual Meeting, 1987–1991, 2012.
Nowakowski, C., and Sharafsaleh, M.A. Preliminary evaluation of drivers' responses to a roadside animal warning system. Proceedings of the Human Factors and Ergonomics Society 57th Annual Meeting, 1953–1957, 2013.
Olson, P.L., and Sivak, M. Perception-response time to unexpected roadway hazards. *Human Factors* 28(1): 91–96, 1986.
Petermeijer, S.M., Abbink, D.A., and de Winter, J.C.F. Should drivers be operating within an automation-free bandwidth? Evaluating haptic steering support systems with different levels of authority. *Human Factors* 57(1): 5–20, 2014.
Popp, M.M., and Faerber, B. Feedback modality for nontransparent driver control actions: Why not visually? In A.G. Gale, J.D. Brown, C.M. Haslegrave, H.W. Kruysse, and S.P. Taylor (Eds). *Vision in Vehicles – IV* (pp. 263–270). Amsterdam: North-Holland, 1993.

Read, J.M., and Saleem, J.J. Task performance and situation awareness with a virtual reality head-mounted display. Proceedings of the Human Factors and Ergonomics Annual Meeting, 2105–2109, 2017.

Reagan, I.J., Bliss, J.P., Van Houten, R., and Hilton, B.W. The effects of external motivation and real-time automated feedback on speeding behavior in a naturalistic setting. *Human Factors* 55(1): 218–230, 2013.

Reed, M.P., and Green, P.A. Comparison of driving performance on-road and in a low-cost simulator using a concurrent telephone dialing task. *Ergonomics* 42(8): 1015–1037, 1999.

Reimer, B., Mehler, B., and Coughlin, J.F. A field study on the impact of variations in short-term memory demands on drivers' visual attention and driving performance across three age groups. *Human Factors* 54(3): 454–468, 2012.

Richter, R.L., and Hyman, W.A. Research note: Driver's brake reaction times with adaptive controls. *Human Factors* 16(1): 87–88, 1974.

Roge, J. Spatial reference frames and driver performance. *Ergonomics* 39(9): 1134–1145, 1996.

Rogers, S.B., and Wierwille, W.W. The occurrence of accelerator and brake pedal actuation errors during simulated driving. *Human Factors* 30(1): 71–81, 1988.

Salmon, P.M., Young, K.L., and Lenne, M.G. Investigating the role of roadway environment in driving errors: An on road study. Proceedings of the Human Factors and Ergonomics Society 55th Annual Meeting, 1879–1883, 2011.

Samost, A., Perlman, D., Domel, A.G., Reimer, B., Mehler, B., Mehler, A., Dobres, J., and McWilliams, T. Comparing the relative impact of Smartwatch and Smartphone use while driving on workload, attention, and driving performance. Proceedings of the Human Factors and Ergonomics Society 59th Annual Meeting, 1602–1606, 2015.

Schrauf, M., Sonnleitner, A., Simon, M., and Kinces, W.E. EEG alpha spindles as indicators for prolonged brake reaction time during auditory secondary tasks in a real road driving study. Proceedings of the Human Factors and Ergonomics Society 55th Annual Meeting, 217–221, 2011.

Schweitzer, N., Apter, Y., Ben-David, G., Lieberman, D.G., and Parush, A. A field study on braking responses during driving. II. Minimum driver braking times. *Ergonomics* 38(9): 1903–1910, 1995.

Shen, S., and Neyens, D.M. Assessing drivers' performance when automated driver support systems fail with different levels of automation. Proceedings of the Human Factors and Ergonomics Society 58th Annual Meeting, 2068–2072, 2014.

Shinar, D., and Stiebel, J. The effectiveness of stationary versus moving police vehicles on compliance with speed limit. *Human Factors* 28(3): 365–371, 1986.

Sidaway, B., Fairweather, M., Sekiya, H., and McNitt-Gray, J. Time-to-collision estimation in a simulated driving task. *Human Factors* 38(1): 101–113, 1996.

Sivak, M., Flannagan, M.J., Sato, T., Traube, E.C., and Aoki, M. Reaction times to neon, LED, and fast incandescent brake lamps. *Ergonomics* 37(6): 989–994, 1994.

Sivak, M., Post, D.V., Olson, P.L., and Donohoe, R.J. Brake responses of unsuspecting drivers to high-mounted brake lights. Proceedings of the Human Factors Society, 139–142, 1980.

Skottle, E.M., Debus, G., Wang, L., and Huestegge, L. Carryover effects of highly automated convoy driving on subsequent manual driving performance. *Human Factors* 56(7): 1272–1283, 2014.

Snyder, H.L. Braking movement time and accelerator-brake separation. *Human Factors* 18(2): 201–204, 1976.
Society of Automotive Engineers Operational definitions of driving performance measures and statistics (Standard J2944). June 30, 2015.
Soliday, S.M. Lane position maintenance by automobile drivers on two types of highway. *Ergonomics* 18(2): 175–183, 1975.
Stahl, P., Donmez, B., and Jamieson, G.A. Correlations among self-reported driving characteristics and simulated driving performance measures. Proceedings of the Human Factors and Ergonomics Society 58th Annual Meeting, 2018–2022, 2014.
Stanton, N., Lew, R., Boyle, N., Dyre, B.P., and Bustmante, E.A. An implementation of a Graded Deceleration Display in a brake light warning system. Proceedings of the Human Factors and Ergonomics Society 55th Annual Meeting, 1573–1577, 2011.
Steyvers, F.J.J.M., and de Waard, D. Road-edge delineation in rural areas: Effects on driving behavior. *Ergonomics* 43(2): 223–238, 2000.
Summala, H. Drivers' steering reaction to a light stimulus on a dark road. *Ergonomics* 24(2): 125–131, 1981.
Summala, H., Nieminen, T., and Punto, M. Maintaining lane position with peripheral vision during in-vehicle tasks. *Human Factors* 38(3): 442–451, 1996.
Svenson, O. Experience of mean speed related to speeds over parts of a trip. *Ergonomics* 19(1): 11–20, 1976.
Szlyk, J.P., Seiple, W., and Viana, M. Relative effects of age and compromised vision on driving performance. *Human Factors* 37(2): 430–436, 1995.
van Winsum, W. Speed choice and steering behavior in curve driving. *Human Factors* 38(3): 434–441, 1996.
van Winsum, W., and Heino, A. Choice of time-headway in car-following and the role of time-to-collision information in braking. *Ergonomics* 39(4): 579–592, 1996.
Vernoy, M.W., and Tomerlin, J. Pedal error and misperceived centerline in eight different automobiles. *Human Factors* 31(4): 369–375, 1989.
Wemeke, J., Kleen, A., and Vollrath, M. Perfect timing: Urgency, not driving situations, influence the best timing to activate warnings. *Human Factors* 56(2): 249–259, 2014.
Wilson, W.T., and Anderson, J.M. The effects of tyre type on driving speed and presumed risk taking. *Ergonomics* 23(3): 223–235, 1980.
Wooller, J. The measurement of driver performance. *Ergonomics* 15(1): 81–87, 1972.
Wu, J., Yang, Y., and Yoshitake, M. Pedal errors among younger and older individuals during different pedal operating conditions. *Human Factors* 56(4): 621–630, 2014.
Xiong, H., Boyle, L.N., Moeckli, J., Dow, B.R., and Brown, T.L. Use patterns among early adopters of adaptive cruise control. *Human Factors* 54(5): 722–733, 2012.
Xiong, Y., and Zhao, G. Taxi-hailing apps: Negative impacts on taxi driver performance. Proceedings of the Human Factors and Ergonomics Society 60th Annual Meeting, 1950–1994, 2016.
Yager, C.E., Cooper, J.M., and Chrysler, S.T. The effects of reading and writing text-based messages while driving. Proceedings of the Human Factors and Ergonomics Society 56th Annual Meeting, 2196–2200, 2012.

2.4.6 Eastman Kodak Company Measures for Handling Tasks

General description – The Eastman Kodak Company (1986) developed a total of eight measures to assess human performance in repetitive assembly, packing, or handling tasks. These eight measures have been divided into "(1) measures of productivity over the shift: total units per shift at different levels and durations of effort and/or exposure, units per hour compared to a standard, amount of time on arbitrary work breaks or secondary work, amount of waste, and work interruptions, distractions, and accidents and (2) quality of output: missed defects/communications, improper actions, and incomplete work" (p. 104).

Strengths and limitations – These measures are well suited for repetitive overt tasks but may be inappropriate for maintenance or monitoring tasks.

Data requirements – Task must require observable behavior.

Thresholds – Not stated.

Source

Eastman Kodak Company. *Ergonomic Design for People at Work*. New York: Van Nostrand Reinhold, 1986.

2.4.7 Haworth-Newman Avionics Display Readability Scale

General description – The Haworth-Newman Avionics Display Readability Scale (see Figure 2.1) is based on the Cooper-Harper Rating Scale. As such, it has a three-level deep branching that systematically leads to a rating of 1 (excellent) to 10 (major deficiencies).

Strengths and limitations – The scale is easy to use. It has been validated in a limited systematic degradation (i.e., masking) of avionics symbology (Chiappetti, 1994). Recommendations from that validation study include: (1) provide a more precise definition of readability, (2) validate the scale using better trained participants, (3) use more realistic displays, and (4) use display resolution, symbol luminance, and symbol size to improve readability.

Data requirements – Participants must have a copy of the scale in front of them during rating.

Thresholds – 1 (excellent) to 10 (major deficiencies).

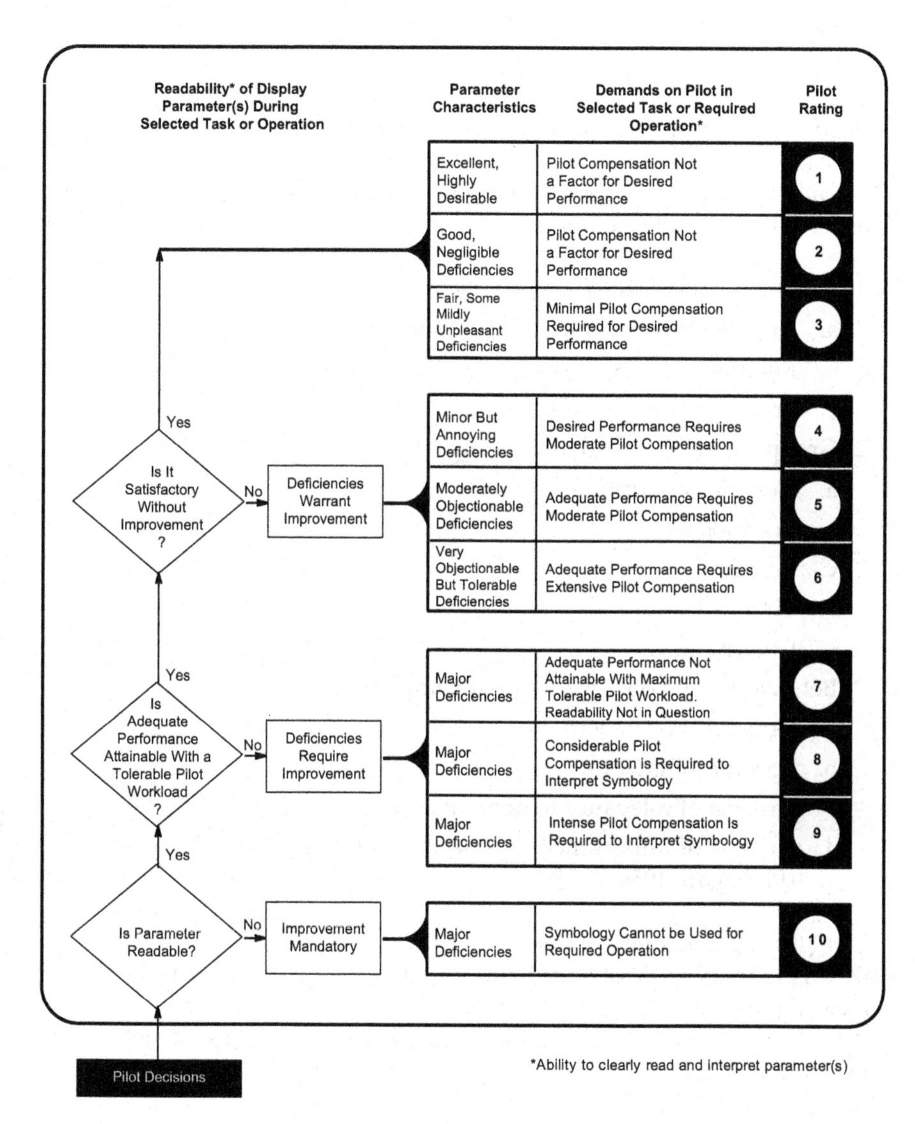

FIGURE 2.1
Haworth-Newman Display Readability Rating Scale (from Haworth, 1993 cited in Chiappetti, 1994).

Source

Chiappetti, C.F. Evaluation of the Haworth-Newman avionics display readability scale. Thesis, Naval Postgraduate School, Monterey, California; September 1994.

2.5 Critical Incidents

The fifth category is critical incidents which are typically used to assess worst case performance.

General description – The Critical Incident Technique includes a set of specifications for collecting data from observed behaviors. These specifications include:

1. Persons to make the observations must have:
   a. Knowledge concerning the activity.
   b. Relation to those observed.
   c. Training requirements.
2. Groups to be observed, including:
   a. General description.
   b. Location.
   c. Persons.
   d. Times.
   e. Conditions.
3. Behaviors to be observed with an emphasis on:
   a. General type of activity.
   b. Specific behaviors.
   c. Criteria of relevance to general aim.
   d. Criteria of importance to general aim (critical prints).
   (Flanagan, 1954, p. 339)

Strengths and limitations – The technique has been used successfully since 1947. It is extremely flexible but can be applied only to observable performance activities.

Data requirements – The specifications listed in the general description paragraph must be applied.

Thresholds – Not stated.

Source

Flanagan, J.C. The critical incident technique. *Psychological Bulletin* 51(4): 327–358, 1954.

2.6 Team Performance Measures

The final category of performance measures is team performance measures. These assess the abilities of two or more persons working in unison to accomplish a task or tasks. These measures assume that human performance varies when part of a team.

Brannick et al. (1997) edited a book that provides an overview of the theory and lessons learned in the measurement of team performance. In that book, Tesluk et al. (1997) identify issues of team work flow (pooled/additive, sequential, reciprocal, intensive) and sources of information (incumbents, subordinates, peers, supervisors or managers, and external experts). The latter can be obtained in the form of surveys, observations, interviews, and archives. These authors also identified issues affecting team performance (selection, training, work design, motivation, and leadership). Marshall et al. (2015) presented what they called the nine Cs of teamwork: (1) cooperation, (2) conflict, (3) coordination, (4) communication, (5) coaching, (6) cognition, (7) composition, (8) culture, and (9) context.

There are literally hundreds of team performance measures. Most of them have been developed for production teams. Examples include: (1) defect percentage, (2) number of accidents, (3) difference between budgeted and actual costs, and (4) customer satisfaction. There are books to help develop measures for teams (e.g., Jones and Schilling, 2000). Jones and Schilling (2000) present eight principles for measuring team performance: capture the team strategy, align the strategy with the organization, stimulate problem solving that results in improved performance, use measurement to focus team meeting, measure the critical items, ensure that team members understand the measures, involve customers in development of the measures, and address the work of each member.

In another book, Heinemann and Zeiss (2002) identify obstacles to measuring team performance in health care settings: (1) many measures are proprietary and require contracting for training to use them, (2) measures have not been reported in a standardized manner enabling easy comparison among them, (3) many measures were developed for other business and are not relevant to health care, and (4) information on the measures is scattered across literature from different disciplines.

Finally, for a review of computerized team performance measures to evaluate military task performance, see Lawson et al. (2012).

Sources

Brannick, M.T., Salas, E., and Prince, C. Team Performance Assessment and Measurement Theory, Methods, and Applications. Mahwah, New Jersey: Lawrence Erlbaum Associates, 1997.

Heinemann, G.D., and Zeiss, A.M. *Team Performance in Health Care Assessment and Development*. New York, New York: Kluwer Academic/Plenum Publishers, 2002.

Jones, S.D., and Schilling, D.J.M. *Measuring Team Performance*. San Francisco, California: Jossey-Bass, 2000.

Lawson, B.D., Kelley, A.M., and Athy, J.R. *A Review of Computerized Team Performance Measures to Identify Military-Relevant, Low-to-Medium Fidelity Tests of Small Group Effectiveness during Shared Information Processing (USAARL 2012-11)*. Fort Detrick, Maryland: United States Army Medical Research and Material Command, May 2012.

Marshall, A., Tisbey, T., and Salas, E. Teamwork and team performance measurement. In J.R. Wilson and S. Sharples (Eds.) *Evaluation of Human Work* (pp. 515–548). Boca Raton: CRC Press, 2015.

Tesluk, P., Mathieu, J.E., and Zaccaro, S.J. Task and aggregation issues in the analysis and assessment of team performance. In M.T. Brannick, E. Salas, and C. Prince (Eds.) *Team Performance Assessment and Measurement Theory, Methods, and Applications* (pp. 197–224). Mahwah, New Jersey: Lawrence Erlbaum Associates, 1997.

2.6.1 Cicek, Koksal, and Ozdemirel's Team Performance Measurement Model

General description – The Team Performance Measurement Model has four categories of team performance measures: (1) structure, (2) process, (3) output, and (4) input. Using the behaviors of effective total quality management teams identified by Champion et al. (1993), O'Brien and Walley (1994), and Storey (1989), Cicek et al. (2005) listed measures for each of the four categories. Structure measurement consists of team members rating their team on a scale of 0 to 100 on behaviors listed in Cicek (1997) relating to clear objectives, communication and conflict management, participation and relationships, good decision making and involvement, knowledge and skills, culture, motivation, and administration. Process measurement consists of data from critical operations and key intermediate results. Output measurements include customer satisfaction and objective measures based on customer expectations, e.g., service times. Input measures are quality of raw materials, information, equipment, services, and management support.

Strengths and limitations – The items on the questionnaires for behaviors were reliable in a hospital setting (Cicek, 1997). The model was applied in a Neurological Sciences Team at a hospital and reported to be effective (Cicek et al., 2005).

Data requirements – List of behaviors from Cicek (1997), flowcharts of process used by team being evaluated, and input and output data.

Thresholds – Not stated.

Sources

Champion, M.A., Medsker, G.J., and Higgs, A.C. Relations between workgroup characteristics and effectiveness: Implications for designing effective workgroups. *Personnel Psychology* 46(3): 823–850, 1993.

Cicek, M.C. A model for performance measurement of total quality teams and an application in a private hospital. Master of Science Thesis, Middle East Technical University, Ankara, 1997.

Cicek, M.C., Koksal, G., and Ozdemirel, N.E. A team performance measurement model for continuous improvement. *Total Quality Management* 16(3): 331–349, 2005.

O'Brien, P., and Walley, P. Total quality teamworking: What's different? *Total Quality Management* 5(3): 151–160, 1994.

Storey, R. Team Building: A Manual for Leaders and Trainers. British Association for Commercial and Industrial Education, 1989.

2.6.2 Collective Practice Assessment Tool

General description – The Collective Practice Assessment Tool (CPAT) was developed by the United Kingdom Ministry of Defense. It is a series of surveys to be completed by Subject Matter Experts (SMEs) evaluating aircrew performance. The surveys include:

1. Planning Phase Assessment to assess leadership, use of information, time management, and decision making
2. Mass Brief Assessment to evaluate briefing clarity and accuracy
3. Assessment Criteria to rate 31 criteria including effectiveness and achieving objectives
4. Mass Debrief Assessment to evaluate briefing clarity and accuracy
5. Training Objectives to rate level of supporting training the objectives
6. Interoperability to assess trust (Gehr et al., 2004)

Strengths and limitations – The CPAT has been used extensively at the Air Warfare Center but it has not been finalized. Further, it is very time consuming to complete and analyze the surveys if they are done manually.

Data requirements – Surveys.

Thresholds – Not stated.

Source

Gehr, S.E., Schreiber, B., and Bennett, W. Within-simulator training effectiveness evaluation. Proceedings of Interservice/Industry Training, Simulation and Education Conference (I/ITSEC), 1652–1661, 2004.

2.6.3 Command and Control Team Performance Measures

General description – Entin et al. (2001) applied the following metrics to assess performance of four diverse teams on a command and control (C2) task: (1) value of enemy assets destroyed, (2) diminished value of own assets, (3) number of friendly assets not destroyed, (4) number of friendly assets lost to hostile fire, (5) number of friendly assets lost to fuel out, (6) number of hostile assets destroyed by friendly action, (7) kill ratio, (8) air refuelings completed, (9) number of transfers of resources in, (10) number of resources out, (11) number of emails sent, and (12) number of emails received.

In another study, Proaps and Bliss (2010) used speed, accuracy, and efficiency to assess team performance of a search task. Speed was the time to find a target, accuracy was finding the correct target, and efficiency was the number of times a team member left the path leading to the target. These authors reported that teams took significantly longer in the difficult than in the easy session. There were no significant differences on accuracy or efficiency.

Proaps and Bliss (2011) evaluated distributed team performance in a computer game requiring two-person teams to find a target among distractors. Difficulty was manipulated by varying the number of distractors and performance by time and error rate.

Strengths and limitations – The results of the analyses of the Entin, Serfaty, Elliott, and Schiflett (2001) study were not presented.

Data requirements – Asset status and email tally.

Thresholds – Not stated.

Sources

Entin, E.B., Serfaty, D., Elliott, L.R., and Schiflett, S.G. *DMT-RNet: An Internet-Based Infrastructure for Distributed Multidisciplinary Investigations of C2 Performance.* Brooks Air Force Base, Texas: Air Force Research Laboratory, 2001.

Proaps, A., and Bliss, J. Team performance as a function of task difficulty in a computer game. Proceedings of the Human Factors and Ergonomics Society 54th Annual Meeting, 2413–2416, 2010.

Proaps, A., and Bliss, J. Distributed team performance in a computer game: The implications of task difficulty. Proceedings of the Human Factors and Ergonomics Society 55th Annual Meeting, 2173–2177, 2011.

2.6.4 Gradesheet

General description – The gradesheet contains 40 items which are rated by self, peers, and instructors. Items are related to air combat and include: radar mechanics, tactics, tactical intercepts, communications, mutual support, and flight leadership. Criteria for rating each of these 40 items are: not applicable, unsafe, lack of ability or knowledge, limited proficiency, recognizes and corrects errors, correct, and unusually high degree of ability.

Strengths and limitations – Krusmark et al. (2004) used data from 148 F-16 pilots grouped into 32 teams performing air combat missions in groups of four aircraft. Ratings were made by seven F-16 subject matter experts using the gradesheet. The average inter-rater reliability was small, 0.42. However, there was very high internal consistency in the ratings, Cronbach's alpha = 98. Ratings were similar across multiple items. The authors performed a principal component analysis to identify the underlying variable accounting for the majority of the variance. One component explained 62.47% of the variance. The next highest component explained only 4.67%.

Data requirements – Ratings of 40 items included in the gradesheet.

Thresholds – Not stated.

Source

Krusmark, M., Schreiber, B.T., and Bennett, W. *The Effectiveness of a Traditional Gradesheet for Measuring Air Combat Team Performance in Simulated Distributed Mission Operations (AFRL-HE-AZ-TR-2004-0090).* Mesa, Arizona: Air Force Research Laboratory, May 2004.

2.6.5 Knowledge, Skills, and Ability

General description – Miller (2001) adapted the Knowledge, Skills, and Ability (KSA) measured by the Teamwork Test. These KSAs are conflict resolution, collaborative problem solving, communication, goal setting and performance management, as well as planning and task management.

Strengths and limitations – Miller (2001) asked 176 undergraduate management majors to work in groups of three to five persons on an organization design task. Each participant completed the Teamwork Test. The team average score and variance were correlated with the group project grade as well as a self-report of satisfaction with the group. There was no significant relationship between team average score on the Teamwork Test and either on task performance or on team satisfaction. There was a trend that teams with high Teamwork Test variances and high Teamwork scores had better task performance ($p = 0.07$).

Data requirements – Complete 35 item Teamwork Test.

Thresholds – Not stated.

Source

Miller, D.L. Examining teamwork KSAs and team performance. *Small Group Research* 32(6): 745–766, 2001.

2.6.6 Latent Semantic Analysis

General description – Latent Semantic Analysis (LSA) is the creation of a word-by-document matrix in which the cells are populated with the frequency of occurrence of that word in that document. Log-entropy term weighting is then applied. Afterwards a singular value decomposition technique is used to identify the significant vectors. LSA does not consider word order or syntax.

Strengths and limitations – Dong et al. (2004) applied LSA to documents produced by collaborative design teams over the 15 weeks of a graduate course in product design. There were eight teams. The Spearman rank correlation coefficient was calculated between the rankings of the semantic coherence and the ranking of the team performance by faculty members. There was a significant positive correlation between the two sets of rankings. The correlation was highest with documents only and lower with documents and email included in the LSA.

Data requirements – Documents produced by the teams.

Thresholds – Not stated.

Source

Dong, A., Hill, A.W., and Agogino, A.M. A document analysis method for characterizing design team performance. *Transactions of the ASME Journal of Mechanical Design* 126(3): 378–385, 2004.

2.6.7 Load of the Bottleneck Worker

General description – Slomp and Molleman (2002) defined the load of the bottleneck worker as "the time in which all the work can be completed by the workers. The worker with the heaviest workload determines this load" (p. 1197). They used the following equation to calculate the load of the bottleneck worker (WB): $WB = max_j \sum_i x_{ij}$ (p. 1200) where j is the worker index, i is the task index, and x_{ij} is the "time (in time units) assigned to worker j to perform task I" (p. 1197).

Strengths and limitations – WB is quantitative and has face validity. It can be used in discrete observable tasks with distant start and completion points. WB was significantly different as a function of training policy, absenteeism, fluctuation in demand, level of cross training, and two of their interactions (training policy by level of cross training and absenteeism by level of cross training; Slomp and Molleman, 2002).

Data requirements – Number of workers, number of tasks, and time for task performance.

Thresholds – Not stated.

Source

Slomp, J., and Molleman, E. Cross-training policies and team performance. *International Journal of Production Research* 40(5): 1193–1219, 2002.

2.6.8 Nieva, Fleishman, and Rieck's Team Dimensions

General description – Nieva et al. (1985) defined five measures of team performance: (1) matching number resources to task requirements, (2) response coordination, (3) activity pacing, (4) priority assignment among tasks, and (5) load balancing.

Strengths and limitations – The five measures are an excellent first step in developing measures of team performance but specific metrics must be developed and tested.

Data requirements – The following group characteristics must be considered when using these measures of team performance: (1) group size, (2) group cohesiveness, (3) intra- and inter-group competition and cooperation, (4) communication, (5) standard communication nets, (6) homogeneity/heterogeneity in personality and attitudes, (7) homogeneity/heterogeneity in ability, (8) power distribution within the group, and (9) group training.

Thresholds – Not stated.

Source

Nieva, V.F., Fleishman, E.A., and Rieck, A. *Team Dimensions: Their Identity, Their Measurement and Their Relationships (Research Note 85-12).* Alexandria, VA: Army Research Institute for the Behavioral and Social Sciences, January 1985.

2.6.9 Project Value Chain

General description – Bourgault et al. (2002) identified the need for performance measures to evaluate virtual organizations of distributed teams. They identified project value chains as critical in this measurement. They define these chains as "the process by which a series of activities are linked together for the purpose of creating value for a client" (p. 3).

Strengths and limitations – Project Value Chains have two distinct advantages. First, they help the team identify "where their input generates values for the end user" (p. 3). Second, Project Value Chains can be linked to other chains including those outside the current team or organization providing increased flexibility.

Data requirements – Process decomposition into activities and links among those activities.

Thresholds – Not stated.

Source

Bourgault, M., Lefebvre, E., Lefebvre, L.A., Pellerin, R., and Elia, E. Discussion of metrics for distributed project management: Preliminary findings. Proceedings of the 35th Hawaii International Conference on System Sciences, 2002.

2.6.10 Targeted Acceptable Responses to Generated Events or Tasks

General description – Targeted Acceptable Responses to Generated Events or Tasks (TARGETs) is an event-based measurement technique (Dwyer et al., 1997). Events required for training and the associated behaviors are identified by Subject Matter Experts. These behaviors are scored as present or absent during an observation procedure. Observers are trained to determine the acceptability of the behaviors observed. Observation forms are tailored to each training exercise or scenario. Scores are based on predetermined observations.

Strengths and limitations – TARGETs has been used in the evaluation of military cargo helicopter training (Fowlkes et al., 1994). The inter-observer reliability was high (+0.94), as was the internal consistency (+0.97), and discriminability between trained and untrained teams.

Data requirements – Knowledge of events and behaviors expected by each member of a team.

Thresholds – Not stated.

Sources

Dwyer, D.J., Fowlkes, J.E., Oser, R.L., and Lane, N.E. Team measurement in distributed environments: The TARGETs methodology. In M.T. Brannick, E. Salas, and C. Prince (Eds.). *Team Performance Assessment and Measurement Theory, Methods, and Applications* (pp. 137–153). Mahwah, New Jersey: Lawrence Erlbaum Associates, 1997.

Fowlkes, J.E., Lane, N.E., Salas, E., Franz, T., and Oser, R. Improving the measurement of team performance: The TARGETs methodology. *Military Psychology* 6: 47–61, 1994.

2.6.11 Team Communication

General description – Harville et al. (2005) developed a set of communication codes (see Figure 2.2) to assess team performance in a command and control (C2) task. The task was developed using a PC-based system emulating United States Air Force Command and Control (C2) tactical operations.

Svensson and Andersson (2006) used speech acts and communication problems to assess performance of fighter pilot teams. Speech acts were: (1) present activity information, (2) future or combat information, (3) tactics,

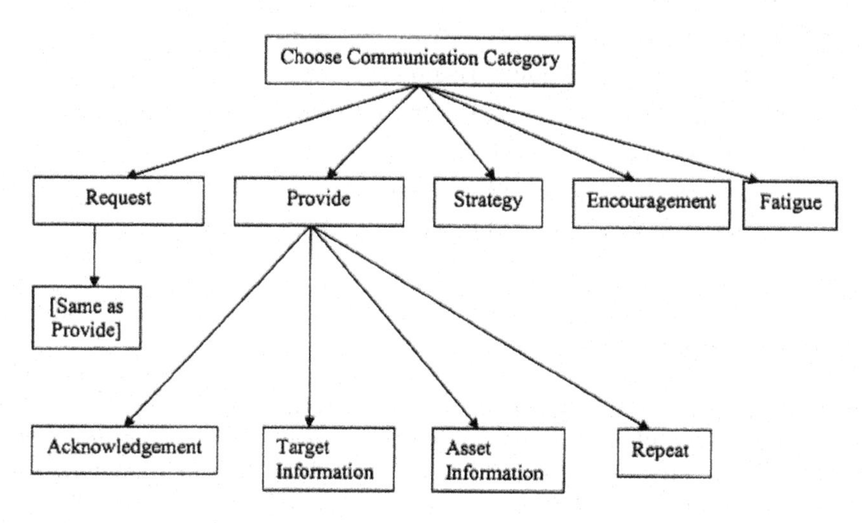

FIGURE 2.2
Communication codes (Harville et al., 2005, p. 7).

(4) communication, (5) question, and (6) other such as practice or identifying issues. Communication problems were categorized: (1) simultaneous, (2) unclear information, (3) uncodeable, (4) unconventional, and (5) uncertain target.

Strengths and limitations. The communications codes were selected for relevance to team performance and distinctiveness. The codes were applied with 95% agreement among coders. Several of the codes were significantly affected by fatigue (i.e., comparison of first and sixth sessions). Total communication and total task communications were significantly less in session 6 than session 1. The same occurred for provide information, provide asset information, request asset information, and strategy. There was a significant increase in communication of fatigue state from the first to sixth session.

Svensson and Andersson (2006) examined the performance of two teams of four pilots and one fighter controller performing aircraft escort or aircraft attack tasks in a simulator. Speech acts frequency was highest when a team was winning. Communication problems were highest when teams were tired.

Hutchins et al. (1999) provided an alternative set of cognitive behaviors of high performing teams based on their review of the literature:

1. "Develop shared understanding of problem, goals, information cues and strategies
2. Anticipate other's resource needs and actions
3. Require little negotiation of what to do and when to do it
4. Volunteer information when needed
5. Provide backup to overworked team members
6. Ensure all members know mission priorities
7. Make and interpret periodic situation updates
8. Provide rationale for decisions
9. Balance workload among team members
10. Use homogeneous, conventional speech patterns"

They measured behaviors 1 through 5 and 10 to evaluate command and control architectures.

Strang et al. (2012) reported no significant effect of cross training or training day on cumulative frequency of team communication. However, there was a significant effect on cumulative duration of team communication. Specifically, the cross trained team had a slightly less duration than noncross-trained teams but only when task demands were high.

TABLE 2.5

Communication Competence Questionnaire

1. The ECP has good command of the language.
2. The ECP medicalizes language appropriately.
3. The ECP is sensitive to others' needs.
4. The ECP typically gets right to the point.
5. The ECP pays attention to what other people say to him/her.
6. The ECP deals with others effectively.
7. The ECP is a good listener.
8. The ECP's writing is understandable.
9. The ECP expresses his/her ideas clearly.
10. The ECP is understandable when he/she speaks.
11. The ECP generally says the right thing at the right time.
12. The ECP is easy to talk to.
13. The ECP usually responds to messages quickly (phone calls, emails, etc.).

Notes: Rated using the following scale: 6, very strong agreement; 5, strong agreement; 4, mild agreement; 3, neutral feelings or don't know; 2, mild disagreement; 1, strong disagreement; 0, very strong disagreement (Cooper et al., 2010, p.14).

In a nonmilitary study, Cooper et al. (2010) applied a communication competence questionnaire to evaluate emergency care practitioners (see Table 2.5).

Data requirements – Records of verbal communications.

Thresholds – Not stated.

Sources

Cooper, S., Endacott, R., and Cant, R. Measuring non-technical skills in medical emergency care: Review of assessment measures. Open Access Emergency Medicine, 7–16, 20 January 2010.

Harville, D., Lopez, N., Elliott, L., and Barnes, C. *Team Communication and Performance during Sustained Working Conditions (AFRL-HE-BR-TR-2005-0085).* Brooks City Base, Texas: Air Force Research Laboratory, May 2005.

Hutchins, S.G., Hocevar, S.P., and Kemple, W.G. Analysis of team communications in "human-in-the-loop" experiments in joint command and control. Presented at the 1999 Command and Control Research and Technology Symposium, Newport, RI, 1999.

Strang, A.J., Funke, G.J., Knott, B.A., Galster, S.M., and Russell, S.M. Effects of cross-training on team performance, communication, and workload in simulated air battle management. Proceedings of the Human Factors and Ergonomics Society 56th Annual Meeting, 1581–1585, 2012.

Svensson, J., and Andersson, J. Speech acts, communication problems, and fighter pilot team performance. *Ergonomics* 49(12–13): 1226–1237, 2006.

2.6.12 Team Effectiveness Measure

General description – Kennedy (2002) developed a Team Effectiveness Measure (TEF) calculated from financial information. TEF is the "ratio of a team's total projects' benefits divided by total cost of implementing and maintain the project" (p. 28). Benefits include increased revenue (annualized) and incremental savings in material, labor, overhead, and other (also annualized). Costs include equipment, material, labor, utilities, and other (also annualized).

In the same year another team effectiveness measure was developed by Hexmoor and Beavers. The Hexmoor and Beavers (2002) measure was based on: (1) efficiency (how much resources are used to achieve goal), (2) influence (how team members affect the performance of other team members), (3) dependence (how much the ability of a team member is affected by the performance of other team members), and (4) redundancy (duplication by two or more team members). All measures are provided in percentages.

Strengths and limitations – TEF was designed to provide an objective measure to make comparisons across teams. Kennedy (2002) applied the measure when comparing 68 teams from two service and five manufacturing companies. Teams varied in size, stage of development, and type. TEF was applied by these diverse teams. A longitudinal study was recommended to further evaluate the utility of these measures of team performance.

The second measure of team effectiveness has not been validated yet. Lum et al. (2011) used a similar metric – number of objectives met. Their task was a simulated military planning task.

Data requirements – Annualized financial data including revenue, material, labor, utilities, and other costs.

Thresholds – Not stated.

Sources

Hexmoor, H., and Beavers, G. Measuring team effectiveness. Proceedings of the International Symposium on Artificial Intelligence and Applications International Conference Applied Infomatics, 351–393, 2002.

Kennedy, F.A. Team performance: Using financial measures to evaluate the influence of support systems on team performance. Dissertation, University of North Texas, Denton, May 2002.

Lum, H.C., Sims, V.K., and Salas, E. Low-level predictors of team performance and success. Proceedings of the Human Factors and Ergonomics Society 55th Annual Meeting, 1457–1461, 2011.

2.6.13 Team Knowledge Measures

General description – Cooke et al. (2003) identified four team knowledge measures: overall accuracy, positional accuracy, accuracy of knowledge of other team members' tasks, and intrateam similarity.

Strengths and limitations – Cooke et al. (2003) reported hands-on cross training of tasks of other team members resulted in significantly better team knowledge than exposure to a conceptual description of the tasks of other team members. The data were collected on 36 three-person teams of undergraduate students. Each team had a dedicated intelligence officer, navigation officer, and pilot. The task was part of a Navy helicopter mission.

Data requirements – Knowledge of components of tasks performed by each member of a team.

Thresholds – Not stated.

Source

Cooke, N.J., Kiekel, P.A., Salas, E., Stout, R., Bowers, C., and Canon-Bowers, J. Measuring team knowledge: A window to the cognitive underpinnings of team performance. Group Dynamics: Theory, Research, and Practice 7(3): 179–199, 2003.

2.6.14 Teamwork Observation Measure

General description – The Teamwork Observation Measure (TOM) measures four dimensions of teamwork: communication, team coordination, situational awareness, and team adaptability (Dwyer et al., 1999). Each of these four dimensions is further divided into factors. Observers then rate on the team's performance on each of these factors from 1 (needs work), 2 (satisfactory), 3 (very good), 4 (outstanding), and not applicable.

Strengths and limitations – TOM has been used to evaluate team performance in aircrew coordination training and naval gunfire support training.

Data requirements – Comments by team members.

Thresholds – Not stated.

Source

Dwyer, D.J., Oser, R.L., Salas, E., and Fowlkes, J.E. Performance measurement in distributed environments: Initial results and implications for training. *Military Psychology* 11(2): 189–215, 1999.

2.6.15 Temkin-Greener, Gross, Kunitz, and Mukamel Model of Team Performance

General description – The Temkin-Greener, Gross, Kunitz, and Mukamel Model of Team Performance was developed in a long-term patient care application.

Strengths and limitations – The model was tested by surveying 26 Program of All-Inclusive Care for the Elderly (PACE) efforts covering 1,860 part-time and full-time employees. The survey included nine items on leadership, 10 each on communication and conflict management, six on coordination, and seven each on team cohesion and perceived unit effectiveness. The response rate was 65%. The internal consistency of responses was +0.7. The authors argued that construct validity was demonstrated by team process variables accounting for 55% of the team cohesion portion and 52% of the team effectiveness of the model. There were also statistically significant differences between professional and paraprofessionals.

Data requirements – Ratings by team members.

Thresholds – Not stated.

Source

Temkin-Greener, H., Gross, D., Kunitz, S.J., and Mukamel, D. Measuring interdisciplinary team performance in a long-term care setting. *Medical Care* 42(5): 472–481, 2004.

2.6.16 Uninhabited Aerial Vehicle Team Performance Score

General description – Cooke et al. (2001) identified the following measures for assessing performance of teams flying reconnaissance Uninhabited Aerial Vehicles (UAVs): film and fuel used, number and type of photographic errors, route deviations, time spent in warning and alarm states, and waypoints visited. They generated a composite team score by subtracting from a starting team score of 1,000 for film and fuel used, unphotographed targets, seconds in alarms state, and unvisited critical waypoints. Additional measures were related to communication and included: number of communication episodes and length of each episode. Finally, the authors presented pairs of tasks and asked the participants to rate the relatedness of concepts.

Strengths and limitations – Cooke et al. (2001) used the data from 11 teams of Air Force Reserve Officer Training Corps cadets for the first study and 18 teams of similar cadets for the second study. All teams performed a UAV reconnaissance task. Relatedness ratings were the best prediction of team performance. In a similar study, Fouse et al. (2012) used a composite score to evaluate the effects of team controlling heterogeneous versus homogenous sets of four versus eight unmanned underwater vehicles (UUV). Teams with fewer vehicles performed better. The composite score included: "number of mines found, targets identified, sonar analyses conducted, duplicate targets found, duplicate mines identified, and times the UUVs were stuck" (p. 398). Teams with homogenous fleets of vehicles did significantly better than teams with heterogeneous fleets. McKendrick et al. (2014) reported significantly poorer team scores without an automated aid.

Data requirements – Data needed include film and fuel used, unphotographed targets, seconds in alarms state, and unvisited critical waypoints. Also, ratings of concept relatedness for task relevant concepts.

Thresholds – Not stated.

Sources

Cooke, N.J., Shope, S.M., and Kiekel, P.A. *Shared-Knowledge and Team Performance: A Cognitive Engineering Approach to Measurement (AFRL-SB-BL-TR-01-0370).* Arlington, Virginia: Air Force Office of Scientific Research, March 29, 2001.

Fouse, S., Champion, M., and Cooke, N.J. The effects of vehicle number and function on performance and workload in human-robot teaming. Proceedings of the Human Factors and Ergonomics Society 56th Annual Meeting, 398–402, 2012.

McKendrick, R., Shaw, T., de Visser, E., Saqer, H., Kidwell, B., and Parasursman, R. Team performance in networked supervisory control of unmanned air vehicles: Effects of automation, working memory, and communication content. *Human Factors* 56(3): 463–475, 2014.

3

Measures of Situational Awareness

Situational Awareness (SA) is knowledge relevant to the task being performed. For example, pilots must know the state of their aircraft, the environment through which they are flying, and relationships between them, such as thunderstorms are associated with turbulence. It is a critical component of decision making and has been included in several models of decision making (e.g., Dorfel and Distelmaier model, 1997; see Figure 3.1). SA has three levels (Endsley, 1991): Level 1, perception of the elements in the environment; Level 2, comprehension of the current situation; and Level 3, projection of future status.

There are four types of SA measures: performance (also known as query methods, Durso and Gronlund, 1999), subjective ratings, simulation (also known as modeling, Golightly, 2015), and physiological measures. Individual descriptions of the first three types of measures of SA are provided in the following sections. Articles describing physiological measures of SA were written by French et al. (2003) and Vidulich et al. (1994). A flowchart to help select the most appropriate measure is given in Figure 3.2.

Note, another categorization of measures of SA is presented in Stanton et al. (2005). Their categories are: SA requirements analysis, freeze probe, real-time probe, self-rating probe, observer rating, and distributed SA. This team has also evaluated 17 SA measures for application in command, control, communication, computers, and intelligence (C4i) applications (Salmon et al., 2006). Their criteria were type of method SA requirements (analysis, freeze probe, real-time probe, self-rating, observer rating, performance measures, eye tracker), domain (air traffic control, civilian aviation, generic, military aviation, military infantry operations, nuclear power), team, subject matter experts required, training time, application time, tools needed, validation studies, advantages, and disadvantages. The authors concluded that all 17 measures were inadequate for the C4i application and recommended combining multiple methods such as performance measures, freeze probe, post-trial self-rating, and observer rating.

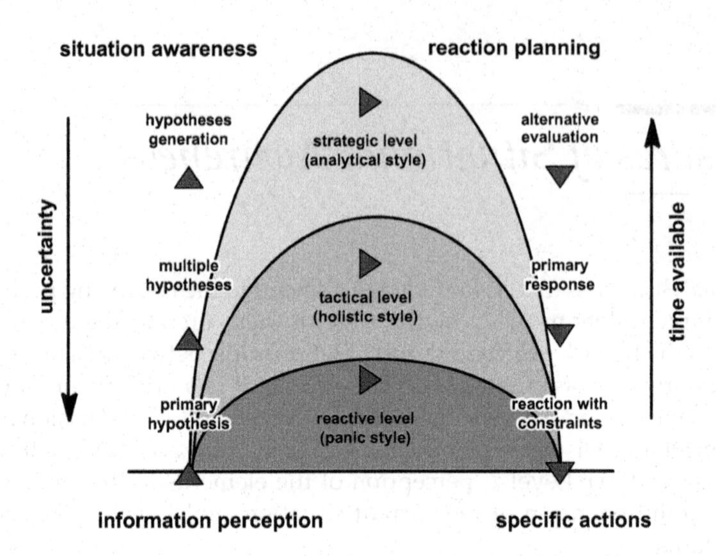

FIGURE 3.1
Decision making under uncertainty and time pressure (Dorfel and Distelmaier, 1997, p. 2).

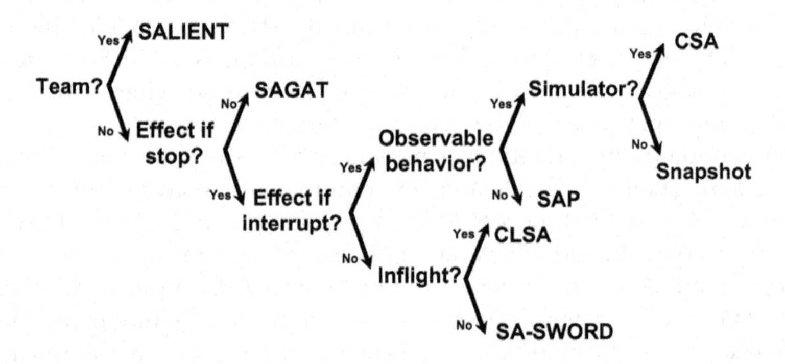

FIGURE 3.2
Guide to selecting a SA measure.

Sources

Dorfel, G., and Distelmaier, H. Enhancing situational awareness by knowledge-based user interfaces. Proceedings of the 2nd Annual Symposium and Exhibition on Situational Awareness in the Tactical Air Environment, 197–205, 1997.

Durso, F.T., and Gronlund, S.D. Situation awareness. In F.T. Durso, R.S. Nickerson, R.W. Schvaneveldt, S.T. Dumais, D.S. Lindsay, and M.T.H. Chi (Eds.) *Handbook of Applied Cognition* (pp. 283–314). New York: John Wiley and Sons, 1999.

Endsley, M.R. Situation awareness in dynamic systems. In R.M. Taylor (Ed.) *Situational Awareness in Dynamic Systems (IAM Report 708)*. Farnborough, UK: Royal Air Force Institute of Aviation Medicine, 1991.
French, H.T., Clark, E., Pomeroy, D., Clarke, C.R., and Seymour, M. Psychophysiological measures of situation awareness. Proceedings of the Human Factors of Decision Making in Complex Systems, Part 6: Assessment and Measurement, Chapter 27, 291–298, August 2003.
Golightly, D. Situation awareness. In J.R. Wilson and S. Sharples (Eds.) *Evaluation of Human Work* (pp. 515–548). Boca Raton: CRC Press, 2015.
Salmon, P., Stanton, N., Walker, G., and Green, D. Situation awareness measurement: A review of applicability for C4i environments. *Applied Ergonomics* 37(2): 225–238, 2006.
Stanton, N.A., Salmon, P.M., Walker, G.H., Barber, C., and Jenkins, D.P. *Human Factors Methods: A Practical Guide for Engineering and Design*. Gower ebook, December 2005.
Vidulich, M.A., Stratton, M., Crabtree, M., and Wilson, G. Performance-based and physiological measures of situational awareness. *Aviation, Space, and Environmental Medicine* 65(5): A7–A12, 1994.

3.1 Performance Measures of SA

There are 12 performance measures of SA each described in a separate section below.

3.1.1 Cranfield Situation Awareness Scale (Cranfield-SAS)

General description – The Cranfield Situation Awareness Scale (Cranfield-SAS) was designed for flight instructors to rate students' SA in general aircraft handling, navigation, basic instrument flying, airways instrument flying, night flying, and commercial airline flight. It has two forms. The long version requires the instructor to rate from 1 (unacceptable) to 9 (excellent) student pilot knowledge (nine questions); understanding and anticipation of future events (five questions); management of stress, effort, and commitment (three questions); ability to attend, perceive, assimilate, and assess information (four questions); and overall awareness. The short form uses the same nine-point rating scale but uses only one rating per category (i.e., (1) knowledge; (2) understand and anticipate future events; (3) management; (4) capacity to perceive, attend, assimilate, and assess; and (5) overall awareness) (Dennehy, 1997).

Data requirements – Instructors must complete a rating for each question. Ratings can be made in flight or in a simulator and either during or after the flight.

Thresholds – 22 to 198 for the long form, 5 to 45 for the short form.

Source

Dennehy, K. *Cranfield – Situation Awareness Scale Users Manual (COA Report Number 9702).* Bedford, England: Applied Psychology Unit Cranfield University, 1997.

3.1.2 Quantitative Analysis of Situation Awareness

General description – The Quantitative Analysis of Situation Awareness (QASA) is a freeze and query method developed by Edgar and Edgar (2007).

Strengths and limitations – Smith and Jamieson (2012) reported a significant difference in QASA scores between manual and information analysis groups in a Cabin Air Management System simulation.

Data requirements – Develop relevant probe questions.

Thresholds – 0 to total number of probe questions.

Sources

Edgar, G.K., and Edgar, H.E. Using signal detection theory to measure situation awareness. The technique, the tool, the test, the way forward. In J. Noyes and M. Cook (Eds.) *Decision Making in Complex Environments* (pp. 373–385). Aldershot, UK: Ashgate, 2007.

Smith, A.G., and Jamieson, G.A. Level of automation effects on situation awareness and functional specificity on automation reliance. Proceedings of the Human Factors and Ergonomics Society 56th Annual Meeting, 2113–2117, 2012.

3.1.3 Quantitative Analysis of Situational Awareness (QUASA)

General description – The Quantitative Analysis of Situational Awareness (QUASA) combines the accuracy of true/false situational awareness probe questions with the self-rating of the confidence of the answer (very high, high, moderate, low, very low).

Strengths and limitations – McGinness (2004) reported that there were significant differences in SA among five teams performing command and control exercises.

Data requirements – A plot of the participant's confidence against the proportion correct of the probe questions is a calibration curve. SA accuracy is measured as the proportion correct. Perceived accuracy is derived from the five-point rating scale.

Thresholds – 0 to 100 for proportion correct, 1 to 5 for confidence.

Source

McGinness, B. Quantitative Analysis of Situational Awareness (QUASA): Applying signal detection theory to true/false probes and self-ratings. 9th International Command and Control Research and Technology Symposium, 159–178, 2004.

3.1.4 SA ANalysis Tool (SAVANT)

General description – The SA ANalysis Tool (SAVANT) uses data from the Air Traffic Control operational systems to probe SA. SAVANT presents a query for three seconds. Queries are aircraft-pair and sector-based questions.

Strengths and limitations – Willems and Heiney (2002) reported significant differences in the time to respond correctly to SAVANT questions. Specifically, Air Traffic Controllers responded faster on the radar side than on the data side. Further, they responded faster to future SA queries when the task load was low. Presentation of the SAVANT screen causes a blind period for the radar or data screen.

Data requirements – Air Traffic Control Subject Matter Experts are required to complete the SAVANT forms with correct answers.

Thresholds – Time to correct responses have a minimum of 0 ms.

Source

Willems, B., and Heiney, M. *Decision Support Automation Research in the En Route Air Traffic Control Environment (DOT/FAA/CT-TN01/10)*. Atlantic City International Airport, NJ: Federal Aviation Administration William J. Hughes Technical Center, January 2002.

3.1.5 SALSA

General description – The acronym is German for "Measuring Situation Awareness of Area Controllers within the Context of Automation" (Hauss and Eyferth, 2003, p. 422). It combines responses to questions posed at simulation freeze with expert ratings of the relevance of the probe question, uses cued recall to reduce confusion, and limits the probe questions to one aircraft at a single freeze.

Strengths and limitations – Hauss and Eyferth (2003) measured SALSA for 11 air traffic controllers over 45 minutes of simulated traffic. They concluded that SALSA was feasible but recommended the development of a taxonomy of air traffic characteristics.

Data requirements – Probe questions, ability to freeze a simulation, and availability of experts to provide ratings of relevance.

Thresholds – Not stated.

Source

Hauss, Y., and Eyferth, K. Securing future ATM-concepts' safety by measuring situation awareness in ATC. *Aerospace Science and Technology* 7(6): 417–427, 2003.

3.1.6 Shared Awareness Questionnaire

General description – The Shared Awareness Questionnaire requires participants to answer objective questions related to an experiment or exercise. Examples include locations critical to mission success and where these locations were on a map (Prytz et al., 2015). The questionnaire is scored on inter-rater agreement and accuracy.

Strengths and limitations – The Shared Awareness Questionnaire was used in an emergency management exercise (Prytz et al., 2015). Fifty participants completed the questionnaire after the exercise was concluded.

Thresholds – Zero to 100% inter-rater agreement and 0 to 100% accuracy.

Source

Prytz, E., Rybing, J., Jonson, C., Petterson, A., Berggren, P., and Johansson, B. An exploratory study of low-level shared awareness measure using mission-critical locations during an emergency response. Proceedings of the Human Factors and Ergonomics Society 59th Annual Meeting, 1152–1156, 2015.

3.1.7 Situational Awareness Global Assessment Technique (SAGAT)

General description – Among the most well-known measure of SA is the Situational Awareness Global Assessment Technique (SAGAT) (Endsley, 1988b). SAGAT was designed around real-time, human-in-the-loop simulation of a military cockpit but could be generalized to other systems. SAGAT uses a graphical computer program for the rapid presentation of queries and data collection. Using SAGAT, the simulation is stopped at random times and the operators are asked questions to determine their SA at that particular point in time. Participants' answers are compared with the correct answers that have been simultaneously collected in the computer database.

"The comparison of the real and perceived situation provides an objective measure of SA" (Endsley, 1988a, p. 101).

Strengths and limitations – The SAGAT technique has been tested in several studies, which demonstrated: (1) empirical validity (Endsley, 1989, 1990b) – the technique of freezing the simulation did not impact participant performance and participants were able to reliably report SA knowledge for up to six minutes after a freeze without memory decay problems; (2) predictive validity (Endsley, 1990b) – linking SAGAT scores to participant performance; and (3) content validity (Endsley, 1990a) – showing appropriateness of the queries used (for an air-to-air fighter cockpit).

Bolstad (1991) correlated SAGAT with scores from potential SA predictor tests. The participants were Northrop employees with tactical flight experience. The top correlations with SAGAT were: attention sharing tracking level (+0.717), immediate/delayed memory total errors (+0.547), encoding speed (+0.547), dot estimation (+0.415), and Group Embedded Figures Test number correct (+0.385).

Bolstad and Endsley (2003) used Goal-Directed Cognitive Task Analysis to develop SAGAT queries. Jones et al. (2004) extended this work to build a toolkit (Designer's Situation Awareness Toolkit) to help develop SAGAT queries.

SAGAT provides objective measures of SA across all of the operators' SA requirements that can be computed in terms of errors or percent correct and can be treated accordingly. However, Sarter and Woods (1991) suggest that SAGAT does not measure SA but rather measures what pilots can recall. Further, Fracker and Vidulich (1991) identified two major problems with the use of explicit measures of SA, such as SAGAT: (1) decay of information and (2) inaccurate beliefs.

Other researchers have simply asked participants to indicate locations of objects on maps. This technique is sometimes referred to as mini-SAGAT or snapshots. It has been used in a wide variety of military and civilian tasks. For example, firefighters have been asked to mark the location of a fire on a map. SA is measured as the deviation between the actual fire and the indicated location on the map (Artman, 1999). Another example is recall of the positions of traffic in front of a participant's vehicle (Johannsdottir and Herdman, 2010). Still another example was used to assess SA of unmanned system operators to locate targets (McDermott and Fisher, 2013). Tippey et al. (2017) used reaction time (RT) to situation awareness probes to evaluate vibrotactile and embedded text cues in general aviation weather alerting.

Vidulich et al. (1995) used a SAGAT-like memory probe to measure SA on a PC-based flight simulator performance. The SAGAT-like measure was sensitive to differences in information presented during the simulation but not to difficulty, thus enabling the differentiation of SA and workload.

SAGAT has been used in many domains including: aviation, command and control, driving, energy, health care, robotics, and submarine display design.

Aviation. Many of the early uses of SAGAT were in military aviation. Fracker (1989) stated from a simulated air threat study "the present data encourage the use of memory probes to measure situation awareness." Fracker (1991) evaluated SA measures in a simulated combat task. Test-retest correlations were significant for identity accuracy, identity latency, envelope sensitivity, and kill probability but not for location error and avoidance failure. Only identity latency was significantly correlated with kill probability. This correlation was used to assess criterion validity. Construct validity was evaluated from correlations of SA metrics with avoidance failure. Three correlations were significant: (1) identity latency, (2) envelope sensitivity, and (3) kill probability. Two correlations were not significant: (1) location error and (2) identity accuracy. There were three significant correlations among SA metrics: (1) identity accuracy and location error, (2) identity accuracy and identity latency, and (3) envelope sensitivity and latency. A 0.10 alpha was used to determine significance. Three correlations were not significant: (1) identity latency and location error, (2) envelope sensitivity and location error, and (3) envelope sensitivity and identify accuracy.

Crooks et al. (2001) asked 165 undergraduate students to rate their SA using SAGAT for a simulated aircraft identification task. The results were mixed. There were significant differences in SAGAT for corridor, size, and direction but not for speed, range, angle, and Identify Friend or Foe. Further, for range and angle, there was a significant difference in the direction opposite the direction hypothesized.

Endsley (1995) did not find any significant difference in SAGAT scores as a function of time (0 to 400 s) between event and response nor did she see any performance decrements in a piloting task between stopping the simulation and not stopping the simulation. Bolstad et al. (2002, 2003) used SAGAT to measure the effectiveness of time-sharing training on enhancing SA. The participants were 24 pilots. The data were collected using the Microsoft *Flight Sim 2000*. There was a significant increase in SA after the training. Those with training went from 22% correct awareness of wind direction prior to training to 55% after training. Those without training went from 58% preflight to 36% postflight.

Snow and Reising (2000) found no significant correlation between SAGAT and SA-SWORD and further only SA-SWORD showed statistically significant effects of visibility and synthetic terrain type in a simulated flight.

SAGAT has also been used in commercial aviation. Boehm-Davis et al. (2010) found no significant differences in awareness for pilots in using Data Comm than using voice. The participants were 24 transport pilots. The data were collected in a simulator.

Prince et al. (2007) used a SAGAT-like approach to measure team situation awareness. In a high-fidelity flight simulator, team situation awareness was measured by two instructors who had completed a three-day training course. The mean correlation between the two instructors on the 48 situation awareness items was $r = +0.88$. There were also significant correlations between team situation awareness and ratings of problem solving for icing, boost pump, and fire emergencies. In a low-fidelity simulator, team simulation awareness was measured by responses to flight status. Questions included: current altitude and heading; controlling agency; last air traffic control (ATC) call; weather at destination, emergency airport, airport bearing and range; and distance to, clock position of, and altitude of traffic. The data were from 41 military aircrews. There were significant correlations between responses to flight status questions and ratings of problem solving for boost pump and fire emergencies but not for icing.

Vu et al. (2010) reported both percent error and probe latency differences for pilots applying different concepts of operation.

SAGAT has been used in evaluations of Air Traffic Control systems as well. Endsley and Rodgers (1996) evaluated SAGAT scores associated with 15 Air Traffic Scenarios associated with operational errors. Their participants were 20 Air Traffic Control Specialists. Low percentage correct on recall items was associated with assigned clearances complete (23.2%), speed (28%), turn (35.1%), and call sign numeric (38.4%). Endsley et al. (2000) compared SAGAT, on-line probes, Situational Awareness Rating Technique (SART), and observer ratings during an Air Traffic Control simulation using 10 experienced air traffic controllers. Only SAGAT Level 2 SA queries were associated with significant differences between display conditions under test.

In a study in the same domain, Jones and Endsley (2004) reported a significant correlation between SAGAT and real-times probes. Their participants were Air Traffic Controllers. Also in Air Traffic Control, Kaber et al. (2006) reported significant differences in level 1 SA between automation modes with information acquisition resulting in the highest SA and action implementation in the lowest SA.

Sethumandhavan (2011a) used SAGAT to measure SA during a simulated air traffic control task using one of four levels of automation (information acquisition, information analysis, decision and action selection, and action implementation automation) and before and after an automation failure. There was significantly higher SA in the information analysis automation condition than the other three automation conditions. Further, participants had significantly higher SA before the occurrence of an automation failure than after it. This author also reported (2011b) that meta-SA (confidence of a participant's own ability to recall aircraft attributes during an air traffic control simulation) was significantly related to SAGAT scores.

In an Air Traffic Control task, Strybel et al. (2009) reported that accuracy was significantly lower for multiple choice probes than for yes/no or rating probes. Willems and Heiney (2002) used SAGAT to evaluate air traffic control SA to airspace sector-based queries. They reported that correct responses for aircraft placement was only about 20%. Further, there was a decrease in SA as task load increased and was less on the data side than on the radar side. Durso et al. (1999) also reported low percent correct scores for Air Traffic Controllers.

In an unusual UAV study, Draper et al. (2000) reported that haptic cues significantly increased the accuracy of identifying turbulence wind direction for UAV pilots. Fern and Shively (2011) also evaluating UAV operator SA used percent correct responses to SA probe questions. They reported significant differences in SA among some but not all display types.

Command and Control. In a command and control domain experiment, Jones and Endsley (2000) reported that response accuracy to real-time probes was associated with significant differences between scenarios (war and peace) but response time did not. For SAGAT, 11 or 21 queries were significantly different between the two scenarios but the aggregate SAGAT score was not. Overall SART scores were significantly different between the two scenarios. French and Hutchinson (2002) applied SAGAT in a military ground force command and control exercise. There was an increase in situation awareness across time. The authors recommended a broader range of probes for future research. Hall et al. (2010) used latency to Commander's Critical Information Report (CCIR) to compare a baseline and a prototype command and control display. The prototype was associated with significantly shorter response latencies.

Lampton et al. (2006) compared responses to event and location to a SA Measure of Team Communications (SAMTC) derived from ratings of team communication on seeking updates, identifying problems, and preventing errors and an adaptation of the Situation Awareness Behaviorally Anchored Rating Scale (SABARS) to rate from 1 to 10 acquisition and dissemination of SA. The authors concluded that all three SA measures were able to measure SA.

In a review of three experiments (military personnel recovery, military command and control, and combat vehicle maneuvering), Cuevas and Bolstad (2010) evaluated the correlation of the team leader's SA using SAGAT and the team members' SA also using SAGAT. There were significant positive corrections for the military personnel recovery and the combat vehicle maneuvering. The authors suggested that the lack of a significant correlation in the command and control experiment was due to missing data. In a separate analysis of the recovery task, Saner et al. (2010) reported significant predictions of shared SA from experience similarity, workload similarity, organizational hub distance, and communication distance but not for shared knowledge.

Parush and Ma (2012) used the mean number of correct responses to the true/false situation awareness statements to assess a team display for managing forest fire incidents. In the following year, Parush and Rustanjaja (2013) used the same method to assess SA for firefighters working in apartments and office buildings.

Giacobe (2013) used SAGAT to compare SA of novice and experienced cyber analysts using text or graphic information. He also measured workload using NASA TLX and perceived awareness using SART. There were no significant effects of either experience or interface style on SAGAT score.

Driving. Bolstad (2000) used queries related to driving to assess differences in SA ability with driver age.

Energy. Hogg et al. (1995) adapted SAGAT to develop the Situation Awareness Control Room Inventory (SACRI) to evaluate nuclear power plant operator's SA. They completed four simulator studies using SACRI and concluded that it was useful in design assessments when used with time and accuracy measures. Similarly, Lenox et al. (2011) adapted SAGAT for electric power transmission and distribution operations.

Tharanathan et al. (2010) evaluated a processing monitoring display. Level 1 SA was measured as the number of process changes identified by the operator, level 1 by the accuracy of responses to probe questions, and level 3 by the number of warning flags predicted. There were significant differences between displays and scenario complexity for SA levels 1 and 2 measures but only a significant difference for scenario complexity on the level 3 SA measure.

In an evaluation of process control displays, Bowden and Rusnock (2015) reported only one significant effect on SAGAT, specifically, graphic displays were associated with higher SA than numeric displays. There was no significant difference between functionally grouped and spatially mapped information displays.

Health care. Luz et al. (2010) used SAGAT ratings after a simulated Mastoidectomy to evaluate the effectiveness of image-guided navigation during surgery. They reported no significant differences in ratings between completing the procedure manually than with the imagery system. Saetrenik (2012) used freeze probes in an emergency handling training scenario. There was a significant difference in SA (average of the nine probes) across the five teams but no significant difference in SA by the role each team member played.

Robotics. Riley and Strater (2006) used SAGAT to compare SA in four robot control modes and reported no significant differences.

Submarine Display Design. Huf and French (2004) reported SART indicating low understanding of the situation in a Virtual Submarine Environment while SAGAT indicated high understanding as evidences by correct responses. Loft et al. (2015) compared SPAM, SAGAT, ATWIT, NASA TLX, and SART ratings from 117 undergraduates performing three submarine tasks: contact classification, closest point of approach, and emergency surface. SPAM did not significantly correlate with SART but did with ATWIT and NASA TLX.

Data requirements – The proper queries must be identified prior to the start of the experiment. Landry and Yoo (2012), based on a Monte Carlo simulation of responses to SA queries, reported that statistical power reduces from 0.8 for 50 queries to 0.58 for 30 queries, 0.24 for 10 queries, and 0.15 for five queries. Endsley (2000) stated that a detailed analysis of SA requirements is required to develop the queries.

Thresholds – Tolerance limits for acceptable deviance of perceptions from real values on each parameter should be identified prior to the start of the experiment.

Sources

Artman, H. Situation awareness and co-operation within and between hierarchical units in dynamic decision making. *Ergonomics* 42(11): 1404–1417, 1999.

Boehm-Davis, D.A., Gee, S.K., Baker, K., and Medina-Mora, M. Effect of party line loss and delivery format on crew performance and workload. Proceedings of the Human Factors and Ergonomics Society 54th Annual Meeting, 126–130, 2010.

Bolstad, C.A. Age-related factors effecting the perception of essential information during risky driving situations. In D.B. Kaber and M.R. Endsley (Eds.) Proceedings of the First Human Performance, Situation Awareness and Automation: User-Centered Design for the New Millennium, October 15–19, 2000.

Bolstad, C.A. Individual pilot differences related to situation awareness. Proceedings of the Human Factors Society 35th Annual Meeting, 52–56, 1991.

Bolstad, C.A., and Endsley, M.R. Measuring shared and team situation awareness in the Army's future objective force. Proceedings of the Human Factors and Ergonomics Society 47th Annual Meeting, 369–373, 2003.

Bolstad, C.A., Endsley, M.R., Howell, C.D., and Costello, A.M. General aviation pilot training for situation awareness: An evaluation. Proceedings of the Human Factors and Ergonomics Society 46th Annual Meeting, 21–25, 2002.

Bolstad, C.A., Endsley, M.R., Howell, C.D., and Costello, A.M. The effect of time-sharing training on pilot situation awareness. Proceedings of the 12th International Symposium on Aviation Psychology, 140–145, 2003.

Bowden, J.R., and Rusnock, C.F. Evaluation of human machine interface design factors on situation awareness and task performance. Proceedings of the Human Factors and Ergonomics Society 59th Annual Meeting, 1361–1365, 2015.

Crooks, C.L., Hu, C.Y., and Mahan, R.P. Cue utilization and situation awareness during simulated experience. Proceedings of the Human Factors and Ergonomics Society 45th Annual Meeting, 1563–1567, 2001.

Cuevas, H.M., and Bolstad, C.A. Influence of team leaders' situational awareness on their team's situational awareness and performance. Proceedings of the Human Factors and Ergonomics Society 54th Annual Meeting, 309–313, 2010.

Draper, M.H., Ruff, H.A., Repperger, D.W., and Lu, L.G. Multi-sensory interface concepts supporting turbulence detection by UAV controllers. Proceedings of the First Human Performance, Situation Awareness and Automation: User-Centered Design for the New Millennium, 107–112, 2000.

Durso, F.T., Hackworth, C.A., Truitt, T.R., Crutchfield, J., Nikolic, D., and Manning, C.A. *Situation Awareness as a Predictor of Performance in En Route Air Traffic Controllers (DOT/FAA/AM-99/3).* Washington, DC: Office of Aviation Medicine, January 1999.

Endsley, M.R. Design and evaluation for situation awareness enhancement. Proceedings of the 32nd Annual Meeting of the Human Factors Society, 97–101, 1988a.

Endsley, M.R. Situational awareness global assessment technique (SAGAT). Proceedings of the National Aerospace and Electronics Conference, 789–795, 1988b.

Endsley, M.R. A methodology for the objective measurement of pilot situation awareness. Presented at the AGARD Symposium on Situation Awareness in Aerospace Operations, Copenhagen, Denmark, October 1989.

Endsley, M.R. *Situation Awareness in Dynamic Human Decision Making: Theory and Measurement (NORDOC 90-49).* Hawthorne, CA: Northrop Corporation, 1990a.

Endsley, M.R. Predictive utility of an objective measure of situation awareness. Proceedings of the Human Factors Society 34th Annual Meeting, 41–45, 1990b.

Endsley, M.R. Toward a theory of situational awareness in dynamic systems. *Human Factors* 37(1): 32–64, 1995.

Endsley, M.R. Direct measurement of situation awareness: Validity and use of SAGAT. In M.R. Endsley and D.J. Garland (Eds.) *Situation Awareness: Analysis and Measurement* (pp. 131–157). Mahwah, NJ: Lawrence Erlbaum Associates, 2000.

Endsley, M.R., and Rodgers, M.D. Attention distribution and situation awareness in air traffic control. Proceedings of the Human Factors and Ergonomics Society 40th Annual Meeting, 82–85, 1996.

Endsley, M.R., Sollenberger, R., and Stein, S. Situation awareness: A comparison of measures. Proceedings of the First Human Performance, Situation Awareness and Automation: User-Centered Design for the New Millennium, October 15–19, 2000.

Fern, L., and Shively, J. Designing airspace displays to support rapid immersions for UAS handoffs. Proceedings of the Human Factors and Ergonomics Society 55th Annual Meeting, 81–85, 2011.

Fracker, M.L. Attention allocation in situation awareness. Proceedings of the Human Factors Society 33rd Annual Meeting, 1396–1400, 1989.

Fracker, M.L. *Measures of Situation Awareness: An Experimental Evaluation (AL-TR-1991-0127).* OH: Wright-Patterson Air Force Base, October 1991.

Fracker, M.L., and Vidulich, M.A. Measurement of situation awareness: A brief review. In R.M. Taylor (Ed.) *Situational Awareness in Dynamic Systems (IAM Report 708)* (pp. 795–797). Farnborough, UK: Royal Air Force Institute of Aviation Medicine, 1991.

French, H.T., and Hutchinson, A. Measurement of situation awareness in a C4ISR experiment. Proceedings of the 7th International Command and Control Research and Technology Symposium. CCRP, Washington, DC, 2002.

Giacobe, N.A. A picture is worth a thousand words. Proceedings of the Human Factors and Ergonomics Society 57th Annual Meeting, 172–176, 2013.

Hall, D.S., Shattuck, L.G., and Bennett, K.B. Evaluation of an ecological interface designed for military command and control. Proceedings of the Human Factors and Ergonomics Society 54th Annual Meeting, 423–427, 2010.

Hogg, D.N., Folleso, K., Strand-Volden, F., and Torralba, B. Development of a situation awareness measure to evaluate advanced alarm systems in nuclear power plant control rooms. *Ergonomics* 38(11): 2394–2413, 1995.

Huf, S., and French, H.T. Situation awareness in a networked virtual submarine. Proceedings of the Human Factors and Ergonomics Society 48th Annual Meeting, 663–667, 2004.

Johannsdottir, K.R., and Herdman, C.M. The role of working memory in supporting drivers' situation awareness for surrounding traffic. *Human Factors* 52(6): 663–673, 2010.

Jones, D.G., and Endsley, M.R. Can real-time probes provide a valid measure of situation awareness? Proceedings of the First Human Performance, Situation Awareness and Automation: User-Centered Design for the New Millennium, 245–250, 2000.

Jones, D.G., and Endsley, M.R. Use of real-time probes for measuring situation awareness. *The International Journal of Aviation Psychology* 14(4): 343–367, 2004.

Jones, D.G., Endsley, M.R., Bolstad, M., and Estes, G. The designer's situation awareness toolkit: Support for user-centered design. Proceedings of the Human Factors and Ergonomics Society 48th Annual Meeting, 653–657, 2004.

Kaber, D.B., Perry, C.M., Segall, N., McClernon, C.K., and Prinzel, L.J. Situation awareness implication of adaptive automation for information processing in an air traffic control-related task. *International Journal of Industrial Ergonomics* 36(5): 447–462, 2006.

Lampton, D.R., Riley, J.M., Kaber, D.B., Sheik-Nainar, M.A., and Endsley, M.R. Use of immersive virtual environments for measuring and training situation awareness. Presented at the U.S. Army Science Conference Orlando, Florida, November 2006.

Landry, S.J., and Yoo, H. Sampling error and other statistical problems with query-based situation awareness measures. Proceedings of the Human Factors and Ergonomics Society 56th Annual Meeting, 292–296, 2012.

Lenox, M.M., Connors, E.S., and Endsley, M.R. A baseline evaluation of situation awareness for electric power system operation supervisors. Proceedings of the Human Factors and Ergonomics Society 55th Annual Meeting, 2044–2048, 2011.

Loft, S., Bowden, V., Braithwaite, J., Morrell, D.B., Huf, S., and Durso, F.T. Situation awareness measures for simulated submarine track management. *Human Factors* 57(2): 298–310, 2015.

Luz, M., Mueller, S., Strauss, G., Dietz, A., Meixenberger, J., and Manzey, D. Automation in surgery: The impact of navigation-control assistance on the performance, workload and situation awareness of surgeons. Proceedings of the Human Factors and Ergonomics Society 54th Annual Meeting, 889–893, 2010.

McDermott, P.L., and Fisher, A. Methodologies for assessing situation awareness of unmanned system operators. Proceedings of the Human Factors and Ergonomics Society 57th Annual Meeting, 167–171, 2013.

Parush, A., and Ma, C. Team displays work, particularly with communication breakdown: Performance and situation awareness in a simulated forest fire. Proceedings of the Human Factors and Ergonomics Society 56th Annual Meeting, 383–387, 2012.

Parush, A., and Rustanjaja, N. The impact of sudden events and spatial configuration on the benefits of prior information to situation awareness and performance. Proceedings of the Human Factors and Ergonomics Society 57th Annual Meeting, 1395–1399, 2013.

Prince, C., Ellis, J.E., Brannick, M.T., and Salas, E. Measurement of team situation awareness in low experience level aviators. *International Journal of Aviation Psychology* 17(1): 41–57, 2007.

Riley, J.M., and Strater, L.D. Effects of robot control mode on situational awareness and performance in a navigation task. Proceedings of the Human Factors and Ergonomics Society 50th Annual Meeting, 540–544, 2006.

Saetrenik, B. A controlled field study of situation awareness measures and heart rate variability in emergency handling teams. Proceedings of the eHuman Factors and Ergonomics Society 56th Annual Meeting, 2006–2010, 2012.

Saner, L.D., Bolstad, C.A., Gonzalez, C., and Cuevas, H.M. Predicting shared situational awareness in teams; A case of differential SA requirements. Proceedings of the Human Factors and Ergonomics Society 54th Annual Meeting, 314–318, 2010.

Sarter, N.B., and Woods, D.D. Situational awareness: A critical but ill-defined phenomenon. *The International Journal of Aviation Psychology* 1(1): 45–57, 1991.

Sethumandhavan, A. Effects of first automation failure on situation awareness and performance in an air traffic control task. Proceedings of the Human Factors and Ergonomics Society 55th Annual Meeting, 350–354, 2011a.

Sethumandhavan, A. Knowing what you know: The role of meta-situation awareness in predicting situation awareness. Proceedings of the Human Factors and Ergonomics Society 55th Annual Meeting, 360–364, 2011b.

Snow, M.P., and Reising, J.M. Comparison of two situation awareness metrics: SAGAT and SA-SWORD. Proceedings of the IEA 2000/HFES 2000 Congress, vol. 3, pp. 49–52, 2000.

Strybel, T.Z., Minakata, K., Nguyen, J., Pierce, R., and Vu, K.L. Optimizing online situation awareness probes in air traffic management tasks. In M.J. Smith and G. Salvendy (Eds.) *Human Interface, Part II, HCII 2009, LNCS 5618* (pp. 845–854). Berlin: Springer, 2009.

Tharanathan, A., Laberge, J., Bullemer, P., Reising, D.V., and McLain, R. Functional versus schematic overview displays: Impact on operator situation awareness in process monitoring. Proceedings of the Human Factors and Ergonomics Society 54th Annual Meeting, 319–323, 2010.

Tippey, K.G., Roady, T., Rodriguez-Paras, C., Brown, L.J., Rantz, W.G., and Ferris, T.K. General aviation weather alerting: The effectiveness of different visual and tactile display characteristics in supporting weather-related decision making. *The International Journal of Aerospace Psychology* 27(3–4): 121–136, 2017.

Vidulich, M.A., McCoy, A.L., and Crabtree, M.S. The effect of a situation display on memory probe and subjective situational awareness metrics. Proceedings of the 8th International Symposium on Aviation Psychology, 765–768, 1995.

Vu, K.L., Strybel, T.Z., Kraut, J., Bacon, P., Minakata, K., Nguyen, J., Rottermann, A., Battiste, V., and Johnson, W. Pilot and controller workload and situation awareness with three traffic management concepts (Paper 978-1-4244-6618-4). IEEE 29th Digital Avionics Systems Conference, 4.A.5-1–4.A.5-10, 2010.

Willems, B., and Heiney, M. *Decision Support Automation Research in the En Route Air Traffic Control Environment (DOT/FAA/CT-TN01/10)*. Atlantic City International Airport, NJ: Federal Aviation Administration William J. Hughes Technical Center, January 2002.

3.1.8 Situational Awareness Linked Instances Adapted to Novel Tasks

General description – The Situational Awareness Linked Instances Adapted to Novel Tasks (SALIANT) was developed to measure team SA. The SALIANT methodology requires five phases: (1) identify team SA behaviors (see Table 3.1), (2) develop scenarios, (3) define acceptable responses, (4) write a script, and (5) create a structured form with columns for scenarios and responses.

Strengths and limitations – SALIANT has been validated using twenty undergraduate students in a four-hour tabletop helicopter simulation. Inter-rater reliability was $r = +0.94$. There were significant correlations between SALIENT score and communication frequency ($r = +0.74$), between SALIENT score and performance ($r = +0.63$). There were no significant correlations between SALIANT score and the teams' shared mental model ($r = -0.04$). Additional validation data are available in Muniz et al. (1998a) and Bowers

TABLE 3.1

Generic Behavioral Indicators of Team SA (Muniz et al., 1998b)

Demonstrated Awareness of Surrounding Environment
Monitored environment for changes, trends, abnormal conditions
Demonstrated awareness of where he/she was
Recognized Problems
Reported problems
Located potential sources of problem
Demonstrated knowledge of problem consequences
Resolved discrepancies
Noted deviations
Anticipated a Need for Action
Recognized a need for action
Anticipated consequences of actions and decisions
Informed others of actions taken
Monitored actions
Demonstrated Knowledge of Tasks
Demonstrated knowledge of tasks
Exhibited skill time sharing attention among tasks
Monitored workload
Shared workload within station
Answered questions promptly
Demonstrated Awareness of Information
Communicated important information
Confirmed information when possible
Challenged information when doubtful
Re-checked old information
Provided information in advance
Obtained information of what is happening
Demonstrated understanding of complex relationship
Briefed status frequently

et al. (1998). In a large experiment (80 men and 180 women), Fink and Major (2000) compared SART and SALIENT related to a helicopter flight simulation game (Werewolf v Comanche). The authors reported that SALIENT had the better measurement characteristics but SART enabled them to assess the interaction between the game and the operator. Kardos (2003) applied the SALIANT methodology to develop a checklist of behavioral SA indicators during development of Tactics, Techniques, and Procedures (TTPs) for the Australian Defence Science and Technology Organisation (DSTO). She described the difficulty in observing some of the behaviors identified such as "monitor others." She recommended excluding some of the behaviors due to lack of applicability, constancy, and not being observed.

Data requirements – Although the generic behaviors in Table 3.1 can be used, scenarios, responses, scripts, and report forms must be developed for each team task.

Thresholds – Not stated.

Sources

Bowers, C., Weaver, J., Barnett, J., and Stout, R. Empirical validation of the SALIANT methodology. Proceedings of the North Atlantic Treaty Organization Research and Technology Organization Meeting 4 (RTO-MP-4, AC/23(HFM)TP/2), 12-1–12-6, April 1998.

Fink, A.A., and Major, D.A. Measuring situation awareness: A comparison of three techniques. Proceedings of the 1st Human Performance, Situation Awareness and Automation: User-Centered Design for the New Millennium, 2000.

Kardos, M. *Behavioral Situation Awareness Measures and the Use of Decision Support Tools in Exercise Prowling Pegasus.* Edinburgh, South Australia: DSTO Systems Sciences Laboratory, 2003.

Muniz, E.J., Salas, E., Stout, R.J., and Bowers, C.A. The validation of a team situational awareness measure. Proceedings for the 3rd Annual Symposium and Exhibition on Situational Awareness in the Tactical Air Environment. Naval Air Warfare Center Aircraft Division, Patuxent River, MD, 183–190, 1998a.

Muniz, E.J., Stout, R.J., Bowers, C.A., and Salas, E. A methodology for measuring team situational awareness: Situational Awareness Linked Indicators Adapted to Novel Tasks (SALIANT). Proceedings of the NATO Human Factors and Medicine Panel on Collaborative Crew Performance in Complex Systems, 20–24, 1998b.

3.1.9 Situation Present Assessment Method (SPAM)

General description – The Situation Present Assessment Method (SPAM) uses response latency to queries as the measure of SA. The queries are asked when all information necessary to answer the query is present, hence eliminating a memory component. It was developed by Durso et al. (1995).

Strengths and limitations – Strybel et al. (2016) used SPAM to compare the situation awareness of retired Air Traffic Controllers using four different separation assurance and spacing concepts in enroute and transitional sectors. The authors reported that in the enroute sector SA was highest when controllers managed separation assurance and the automation managed the spacing. In the transitional sector, SA was highest when the controllers managed both functions.

Vu et al. (2009) reported students training in Air Traffic Control answered significantly more probe questions than retired Air Traffic Controllers.

Durso et al. (1999) applied SPAM to the comparison of Air Traffic Control scenarios. There were no significant differences. Baker et al. (2012) reported no significant difference in SPAM scores between data comm and voice communication for pilots using a low-fidelity flight simulator. Pierce (2012) reported that SPAM reduced the number of correctly handled aircraft.

In a domain other than Air Traffic Control, Schuster et al. (2012) reported significantly higher SPAM scores (i.e., number of questions answered correctly) for mission 2 than mission 3 with a human-robot team task of identifying soldiers presented in a simulated Military Operations in Urban Terrain (MOUT).

Loft et al. (2015b) reported no significant differences in SPAM accuracy or time to respond to SPAM queries between two levels of task load. Loft et al. (2015a) compared SPAM, SAGAT, ATWIT, NASA TLX, and SART ratings from 117 undergraduates performing three submarine tasks: contact classification, closest point of approach, and emergency surface. SPAM did not significantly correlate with SART but did with ATWIT and NASA TLX.

Data requirements – Questions must be identified as well as start time for measuring response latency.

Thresholds – Not stated.

Sources

Baker, K.M., DiMare, S.K., Nelson, E.T., and Boehm-Davis, D.A. Effect of data communications on pilot situation awareness, decision making, and workload. Proceedings of the Human Factors and Ergonomics Society 56th Annual Meeting, 1787–1788, 2012.

Durso, F.T., Hackworth, C.A., Truitt, T.R., Crutchfield, J., Nikolic, D., and Manning, C.A. *Situation Awareness as a Predictor of Performance in En Route Air Traffic Controllers (DOT/FAA/AM-99/3).* Washington, DC: Office of Aviation Medicine, January 1999.

Durso, F.T., Truitt, T.R., Hackworth, C.A., Crutchfield, J.M., Nikolic, D., Moertl, P.M., Ohrt, D., and Manning, C.A. Expertise and chess: Comparing situation awareness methodologies. Proceedings of the International Conference on Experimental Analysis and Measurement of Situation Awareness, 295–303, 1995.
Loft, S., Bowden, V., Braithwaite, J., Morrell, D.B., Huf, S., and Durso, F.T. Situation awareness measures for simulated submarine track management. *Human Factors* 57(2): 298–310, 2015a.
Loft, S., Sadler, A., Braithwaite, J., and Huff, S. The chronic detrimental impact of interruptions in a simulated submarine track management task. *Human Factors* 57(8): 1417–1426, 2015b.
Pierce, R.S. The effect of SPAM administration during a dynamic simulation. *Human Factors* 54(5): 838–848, 2012.
Schuster, D., Keebler, J.R., Jentsch, F., and Zuniga, J. Comparison of SA measurement techniques in a human-robot team task. Proceedings of the Human Factors Society 56th Annual Meeting, 1713–1717, 2012.
Strybel, T.Z., Vu, K.L., Chiappe, D.L., Morgan, C.A., Morales, G., and Battiste, V. Effects of NextGen concepts of operation for separation assurance and interval management on air traffic controller situation awareness, workload, and performance. *International Journal of Aviation Psychology* 26(1–2): 1–14, 2016.
Vu, K.P.L., Minakata, K., Nguyen, J., Kraut, J., Raza, H., Battiste, V., and Strybel, T.Z. Situation awareness and performance of student versus experienced air traffic controllers. In M.J. Smith and G. Salvendy (Eds.) *Human Interface, Part II, HCII 2009, LNCS5618* (pp. 865–874), 2009.

3.1.10 Tactical Rating of Awareness for Combat Environments (TRACE)

General description – The Tactical Rating for Awareness for Combat Environments (TRACE) estimates SA from the accuracy of responses to periodic situation reports (SITREP) and the elapsed time from the request for the SITREP and the completion of all SITREP line items (Hall, 2009).

Strengths and limitations – Hall et al. (2010) used TRACE to compare a baseline and a prototype command and control display. The prototype was associated with both significantly shorter response latencies and significantly greater number of correct responses.

Data requirements – Questions must be identified as well as start time for measuring response time.

Thresholds – Not stated.

Sources

Hall, D.S. Raptor: An empirical evaluation of an ecological interface designed to increase warfighter cognitive performance. Master's Thesis. Naval Postgraduate School, Monterey, CA, June 2009.

Hall, D.S., Shattuck, L.G., and Bennett, K.B. Evaluation of an ecological interface designed for military command and control. Proceedings of the Human Factors and Ergonomics Society 54th Annual Meeting, 423–427, 2010.

3.1.11 Temporal Awareness

General description – Temporal awareness has been defined as "the ability of the operator to build a representation of the situation including the recent past and the near future" (Grosjean and Terrier, 1999, p. 1443). It has been hypothesized to be critical to process management tasks.

Strengths and limitations – Temporal awareness has been measured as the number of temporal and ordering errors in a production line task, number of periods in which temporal constraints were adhered to, and the temporal landmarks reported by the operator to perform his or her task. Temporal landmarks include relative ordering of production lines and clock and mental representation of the position of production lines.

Data requirements – Correct time and order must be defined with tolerances for error data. Temporal landmarks must be identified during task debriefs.

Thresholds – Not stated.

Source

Grosjean, V., and Terrier, P. Temporal awareness: Pivotal in performance? *Ergonomics* 42(11): 1443–1456, 1999.

3.1.12 Virtual Environment Situation Awareness Rating System

General description – The Virtual Environment Situation Awareness Rating System (VESARS) is a software tool that measures SA in virtual and real-world training environments. It has three metrics: (1) real-time SA queries of individual, shared, and team SA, (2) real-time SA behavior ratings of individual and team SA by subject matter experts, and (3) real-time SA communication ratings also made by subject matter experts (Strater et al., 2013).

Strengths and limitations – The VESARS has been integrated into the Virtual Battlespace 2 training suite. Scielzo et al. (2010) used VESARS to measure the SA of two squadron leaders during a training game. They concluded that VESARS could be successfully used real time if the raters were very familiar with the rating system.

Data requirements – Well-defined queries and well-trained raters.

Thresholds – 0% to 100%.

Sources

Scielzo, S., Davis, F., Riley, J.M., Hyatt, J., Lampton, D., and Merlo, J. Leveraging serious games and advanced training technologies for enhanced cognitive skill development. Proceedings of the Human Factors and Ergonomics Society 54th Annual Meeting, 2408–2412, 2010.

Strater, L., Riley, J., Davis, F., Procci, K., Bowers, C., Beidel, D., Brunnell, B., Proaps, A., Sankaranarayanan, G., Li, B., De, S., and Cao, C.G.L. Me and my VE, part 2. Proceedings of the Human Factors and Ergonomics Society 57th Annual Meeting, 2127–2131, 2013.

3.2 Subjective Measures of SA

Subjective measures of SA share many of the advantages and limitations of subjective measures of workload discussed in Section 3.2. Advantages include: inexpensive, easy to administer, and high face validity. Disadvantages include: inability to measure what the subject cannot describe well in words and requirement for well-defined questions.

3.2.1 China Lake Situational Awareness

General description – The China Lake Situational Awareness (CLSA) is a five-point rating scale (see Table 3.2) based on the Bedford Workload Scale. It was designed at the Naval Air Warfare Center at China Lake to measure SA in flight (Adams, 1998).

Strengths and limitations – Jennings et al. (2004) reported a significant increase in CLSA ratings associated with a Tactile Situational Awareness System that provided tactile cues for maintaining aircraft position. The participants were 11 pilots flying a Bell 205 Helicopter.

Bruce Hunn (personal communication, 2001) argued that the CLSA "fails to follow common practice in rating scale design, does not provide diagnostic results and in general, is unsuitable for assessing SA in a test environment." He argues that the scale cannot measure SA since it does not include the three components of SA: perception, comprehension, or projection. Further, Hunn argues that the terminology is not internally consistent, includes multiple dimensions and compound questions, and has not been validated.

Data requirements – Points in the flight during which the aircrew are asked to rate their SA using the CLSA Rating Scale must not compromise safety.

Thresholds – 1 (very good) to 5 (very poor).

TABLE 3.2
China Lake SA Rating Scale

SA Scale Value	Content
VERY GOOD 1	Full knowledge of a/c energy state/tactical environment/mission; Full ability to anticipate/accommodate trends
GOOD 2	Full knowledge of a/c energy state/tactical environment/mission; Partial ability to anticipate/accommodate trends; No task shedding
ADEQUATE 3	Full knowledge of a/c energy state/tactical environment/mission; Saturated ability to anticipate/accommodate trends; Some shedding of minor tasks
POOR 4	Fair knowledge of a/c energy state/tactical environment/mission; Saturated ability to anticipate/accommodate trends; Shedding of all minor tasks as well as many not essential to flight safety/mission effectiveness
VERY POOR 5	Minimal knowledge of a/c energy state/tactical environment/mission; Oversaturated ability to anticipate/accommodate trends; Shedding of all tasks not absolutely essential to flight safety/mission effectiveness

Sources

Adams, S. Practical considerations for measuring situational awareness. Proceedings for the 3rd Annual Symposium and Exhibition on Situational Awareness in the Tactical Air Environment, 157–164, 1998.

Jennings, S., Craig, G., Cheung, B., Rupert, A., and Schultz, K. Flight-test of a tactile situational awareness system in a land-based deck landing task. Proceedings of the Human Factors and Ergonomics Society 48th Annual Meeting, 142–146, 2004.

3.2.2 Crew Awareness Rating Scale

General description – The Crew Awareness Rating Scale (CARS) has eight scales (see Table 3.3) that are rated from 1 (the ideal case) to 4 (the worst case).

Strengths and limitations – McGuinness and Foy (2000) reported that the CARS had been successfully used in studies of airline pilot eye movements, automation trust on flight decks of commercial aircraft, and military command and control. Foy and McGuinness (2000) reported significant differences between pilots provided with a Traffic Collision Avoidance System (TCAS) and those without – but only for the resolution component. McGuinness et al. (2000) used CARS to compare SA between conventional and digital command and control displays. Understanding the big picture had significantly higher SA in the digital condition. McGuinness and Ebbage (2002) used CARS to evaluate the SA of seven Commanding Officer/Operations Officer teams in the Royal Military College of Science. All seven

TABLE 3.3

Definitions of CARS Rating Scales

Perception – the assimilation of new information
1. The content of perception – is it reliable and accurate?
2. The processing of perception – is it easy to maintain?
Comprehension – the understanding of information in context
3. The content of comprehension – is it reliable and accurate?
4. The processing of comprehension – is it easy to maintain?
Projection – the anticipation of possible future developments
5. The content of projection – is it reliable and accurate?
6. The processing of projection – is it easy to maintain?
Integration – the synthesis of the above with one's course of action
7. The content of integration – is it reliable and accurate?
8. The processing of integration – is it easy to maintain?

teams completed two two-hour land reconnaissance missions; one with a standard radio and one with a digital map and electronic text messaging. Content ratings gradually improved over the exercise. The processing ratings were not significantly different between the standard and digitized versions of the mission. Content, however, was higher in the digital version except for the Commanding Officers who rated comprehension higher using the standard radio.

CARS has also been used to evaluate the relationships among workload, teamwork, SA, and performance (Berggren et al. 2011). They reported that seven of the eight CARS questions were identified in a Principal Component Analysis (PCA) as a latent factor in teamwork. The other factors were performance, workload, and teamwork. The participants were 18 two-person teams of university students fighting a virtual fire.

In a recent aviation study, Stelzer et al. (2013) reported no differences in SA as measured by CARS with and without a block occupancy display. The participants were 11 former or current air traffic controllers.

Data requirements – Use of the standard CARS rating scales.

Sources

Berggren, P., Prytz, E., Johansson, B., and Nahlinder, S. The relationship between workload, teamwork, situation awareness, and performance in teams: A microworld study. Proceedings of the Human Factors and Ergonomics Society 55th Annual Meeting, 851–855, 2011.

Foy, L., and McGuinness, B. Implications of cockpit automation for crew situational awareness. Proceedings of the 1st Human Performance, Situation Awareness and Automation: User-Centered Design for the New Millennium, 101–106, 2000.

McGuinness, B., and Ebbage, L. *Assessing Human Factors in Command and Control: Workload and Situational Awareness Metrics.* Bristol, United Kingdom: BAE Systems Advanced Technology Centre, May 2002.

McGuinness, B., and Foy, L. A subjective measure of SA: The Crew Awareness Rating Scale (CARS). Proceedings of the Human Performance, Situation Awareness, and Automation Conference, 286–291, 2000.

McGuinness, B., Foy, L., and Forsey, T. Battlespace digitization: SA issues for commanders. Proceedings of the First Human Performance, Situation Awareness and Automation: User-Centered Design for the New Millennium, 125, 2000.

Stelzer, E.K.M., Chong, R.S., Stevens, R.K., and Nene, V.D. Controller use of a block occupancy-based surface surveillance display for surface management. Proceedings of the Human Factors and Ergonomics Society 57th Annual Meeting, 51–55, 2013.

3.2.3 Crew Situational Awareness

General description – Mosier and Chidester (1991) developed a method for measuring situational awareness of air transport crews. Expert observers rate crew coordination performance and identify and rate performance errors (type 1, minor errors; type 2, moderately severe errors; and type 3, major, operationally significant errors). The experts then develop information transfer matrices identifying time and source of item requests (prompts) and verbalized responses. Information is then classified into decision or nondecision information.

Strengths and limitations – The method was sensitive to type of errors and decision prompts.

Data requirements – The method requires open and frequent communication among aircrew members. It also requires a team of expert observers to develop the information transfer matrices.

Thresholds – Not stated.

Source

Mosier, K.L., and Chidester, T.R. Situation assessment and situation awareness in a team setting. In R.M. Taylor (Ed.) *Situation Awareness in Dynamic Systems (IAM Report 708).* Farnborough, UK: Royal Air Force Institute of Aviation Medicine, 1991.

3.2.4 Mission Awareness Rating Scale (MARS)

General description – The Mission Awareness Rating Scale (MARS) is an eight-question rating scale designed for use by infantry personnel (see Table 3.4).

Strengths and limitations – Matthews and Beal (2002) asked 16 cadets attending the U.S. Military Academy to provide MARS ratings during a battalion level exercise. Platoon leaders rated their SA higher than squad leaders.

Data requirements – Completion of the ratings for 8 questions.
Thresholds – 0 to 3 for content and workload.

Source

Matthews, M.D., and Beal, S.A. *Assessing Situation Awareness in Field Training Exercises (Research Report 1795).* Alexandria, VA: U.S: Army Research Institute for the Behavioral and Social Sciences, September 2002.

3.2.5 Human Interface Rating and Evaluation System

General description – The Human Interface Rating and Evaluation System (HiRes) is a generic judgment-scaling technique developed by Budescu et al. (1986).

Strengths and limitations – HiRes has been used to evaluate SA (Fracker and Davis, 1990). These authors reported a significant effect on the number of enemy aircraft in a simulation and HiRes rating.

Data requirements – HiRes ratings are scaled to sum to 1.0 across all the conditions to be rated.

Thresholds – 0 to 1.0.

Sources

Budescu, D.V., Zwick, R., and Rapoport, A. A comparison of the Eigen value and the geometric mean procedure for ratio scaling. *Applied Psychological Measurement* 10(1): 69–78, 1986.

Fracker, M.L., and Davis, S.A. Measuring operator situation awareness and mental workload. Proceedings of the 5th Mid-Central Ergonomics/Human Factors Conference, 23–25, 1990.

3.2.6 Situation Awareness for SHAPE

General description – As part of EUROCONTROL's Solution for Human-Automation Partnerships in European ATM (SHAPE) project, the Situation Awareness for SHAPE (SASHA) was developed. There are two forms of SASHA: (1) SASHA on-Line (SASHA_L) is a set of queries and (2) SASHA Questionnaire (SASHA_Q) consists of questions related to elements of SA identified by Air Traffic Controllers (Straeter and Woldring, 2003). For SASHA_L, a subject matter expert views an Air Traffic Controller's

TABLE 3.4

Mission Awareness Rating Scales

Instructions. Please answer the following questions about the mission you just completed. Your answers to these questions are important in helping us evaluate the effectiveness of this training exercise. Check the response that best applies to your experience.

The first four questions deal with your ability to detect and understand important cues present during the mission.

1. Please rate your ability to identify mission-critical cues in this mission.

___ very easy – able to identify all cues
___ fairly easy – could identify most cues
___ somewhat difficult – many cues hard to identify
___ very difficult – had substantial problems identifying most cues

2. How well did you understand what was going on during the mission?

___ very well – fully understood the situation as it unfolded
___ fairly well – understood most aspects of the situation
___ somewhat poorly – had difficulty understanding much of the situation
___ very poorly – the situation did not make sense to me

3. How well could you predict what was about to occur next in the mission?

___ very well – could predict with accuracy what was about to occur
____ fairly well – could make accurate predictions most of the time
___ somewhat poor – misunderstood the situation much of the time
___ very poor – unable to predict what was about to occur

4. How aware were you of how to best achieve your goals during this mission?

___ very aware – knew how to achieve goals at all times
___ fairly aware – knew most of the time how to achieve mission goals
___ somewhat unaware – was not aware of how to achieve some goals
___ very unaware – generally unaware of how to achieve goals

The last four questions ask how difficult it was for you to detect and understand important cues present during the mission.

5. How difficult – in terms of mental effort required – was it for you to identify or detect mission-critical cues in the mission?

___ very easy – could identify relevant cues with little effort
___ fairly easy – could identify relevant cues, but some effort required
___ somewhat difficult – some effort was required to identify most cues
___ very difficult – substantial effort required to identify relevant cues

6. How difficult – in terms of mental effort – was it to understand what was going on during the mission?

___ very easy – understood what was going on with little effort
___ fairly easy – understood events with only moderate effort
___ somewhat difficult – hard to comprehend some aspects of situation
___ very difficult – hard to understand most or all aspects of situation

7. How difficult – in terms of mental effort – was it to predict what was about to happen during the mission?

___ very easy – little or no effort needed
___ fairly easy – moderate effort required

(*Continued*)

TABLE 3.4 (CONTINUED)

Mission Awareness Rating Scales

___ somewhat difficult – many projections required substantial effort
___ very difficult – substantial effort required on most or all projections
8. How difficult – in terms of mental effort – was it to decide on how to best achieve mission goals during this mission?
___ very easy – little or no effort needed
___ fairly easy – moderate effort required
___ somewhat difficult – substantial effort needed on some decisions
___ very difficult – most or all decisions required substantial effort (Matthews and Beal, 2002, p. A-1–A-2).

displays and in real time asks SA-related questions. The questions are rated by the subject matter expert on their *operational* importance (Straeter and Woldring, 2003). The time for an Air Traffic Controller to respond to the question is rated as OK, too long, or too short. An example of a question is "Which aircraft needs to be transferred next?" (Straeter and Woldring, 2003, p. 50). SASHA_Q is a self-rating questionnaire. An example is presented in Figure 3.3.

Strengths and limitations – Dehn (2008) described the steps taken in the development of the SASHA: (1) literature review to obtain initial set of items, (2) requirement-based review for easy administration, ease of understanding, consistent format, and scoring key provided, (3) collection of expert feedback, and (4) initial empirical study. The study participants were 24 active air traffic controllers who completed the SASHA. Items were eliminated from the questionnaire if they reduced internal consistency or were redundant. Jipp and Papenfuss (2011) reported problems replicating the scales as well as finding results contrary to previous research. Their participants were 12 professional tower controllers working in teams during an airspace simulation. The SHAPE Teamwork Questionnaire (STQ) was completed after each of the three simulation runs.

In the previous working period(s) ...

	never	seldom	sometimes	often	more often	very often	always
1) ... I was ahead of the traffic.	0	1	2	3	4	5	6
2) ... I started to focus on a single problem or a specific area of the sector.	0	1	2	3	4	5	6

FIGURE 3.3
SASHA (Dehn, 2008, p. 138).

Data requirements – Data are responses to questionnaire items. A scoring key for SASHA is available from EUROCONTROL Headquarters. It includes scores for both the inverted and noninverted questions. There is also a user's guide (EUROCONTROL Edition Number 0.1, 30 July 2007).

Thresholds – The data are on an ordinal scale and must be treated accordingly when statistical analysis is applied to the data. Non-parametric statistics may be the most appropriate analysis method.

Sources

Dehn, D.M. Assessing the impact of automation on the air traffic controller: The SHAPE questionnaires. *Air Traffic Control Quarterly* 16(2): 127–146, 2008.

EUROCONTROL. *The New SHAPE Questionnaires: A User's Guide.* Edition Number 0.1. 30 July 2007. http://www.eurocontrol.int/humanfactors/public/standard_page/15_newsletter_SHAPE.html

Jipp, M., and Papenfuss, A. Reliability and validity of the SHAPE Teamwork Questionnaire. Proceedings of the Human Factors and Ergonomics Society 55th Annual Meeting, 115–119, 2011.

Straeter, O., and Woldring, M. *The Development of Situation Awareness Measures in ATM Systems (HRS/HSP-005-REP-01).* Brussels: EATMP Infocenter. EUROCONTROL Headquarters, June 27, 2003.

3.2.7 Situation Awareness Behavioral Rating Scale (SABARS)

General description – The Situation Awareness Behavioral Rating Scale (SABARS) is a 28-question rating scale measuring SA of infantry personnel (see Table 3.5). It is completed by observers.

Strengths and limitations – Matthews and Beal (2002) asked six infantry officers and four infantry noncommissioned officers to rate 16 U.S. Military Academy during a battalion level exercise. There were no significant differences in the ratings given by the commissioned and noncommissioned officers. In an earlier study, Matthews et al. (2000) recommended use of three SA measures concurrently: SABARS, SAGAT, and a Participant Situation Awareness Questionnaire (PSAQ).

Data requirements – Completion of the ratings for 28 questions.

Thresholds – 1 to 6 for each question.

TABLE 3.5

Situation Awareness Behavioral Rating Scale

Rating Scale:	
1. Very Poor	
2. Poor	
3. Borderline	
4. Good	
5. Very good	
6. Not applicable/can't say	
RATING (Circle one)	
1. Sets appropriate levels of alert	1 2 3 4 5 6
2. Solicits information from subordinates	1 2 3 4 5 6
3. Solicits information from civilians	1 2 3 4 5 6
4. Solicits information from commanders	1 2 3 4 5 6
5. Effects coordination with other platoon/squad leaders	1 2 3 4 5 6
6. Communicates key information to commander	1 2 3 4 5 6
7. Communicates key information to subordinates	1 2 3 4 5 6
8. Communicates key information to other platoon/squad leaders	1 2 3 4 5 6
9. Monitors company net	1 2 3 4 5 6
10. Assesses information received	1 2 3 4 5 6
11. Asks for pertinent intelligence information	1 2 3 4 5 6
12. Employs squads/fire teams tactically to gather needed information	1 2 3 4 5 6
13. Employs graphic or other control measures for squad execution	1 2 3 4 5 6
14. Communicates to squads/fire teams, situation and commander's intent	1 2 3 4 5 6
15. Utilizes a standard reporting procedure	1 2 3 4 5 6
16. Identifies critical mission tasks to squad/fire team leaders	1 2 3 4 5 6
17. Ensures avenues of approach are covered	1 2 3 4 5 6
18. Locates self at vantage point to observe main effort	1 2 3 4 5 6
19. Deploys troops to maintain platoon/squad communications	1 2 3 4 5 6
20. Uses assets to effectively assess environment	1 2 3 4 5 6
21. Performs a leader's recon to assess terrain and situation	1 2 3 4 5 6
22. Identifies observation points, avenues of approach, key terrain, obstacles, cover and concealment	1 2 3 4 5 6
23. Assesses key finds and unusual events	1 2 3 4 5 6
24. Discerns key/critical information from maps, records, and supporting site information	1 2 3 4 5 6
25. Discerns key/critical information from reports received	1 2 3 4 5 6
26. Projects future possibilities and creates contingency plans	1 2 3 4 5 6
27. Gathers follow up information when needed	1 2 3 4 5 6
28. Overall Situation Awareness Rating	1 2 3 4 5 6

Matthews and Beal (2002, p. B-1).

Sources

Matthews, M.D., and Beal, S.A. *Assessing Situation Awareness in Field Training Exercises (Research Report 1795).* Alexandria, VA: U.S. Army Research Institute for the Behavioral and Social Sciences, September 2002.

Matthews, M.D., Pleban, R.J., Endsley, M.R., and Strater, L.D. Measures of infantry situation awareness for a virtual MOUT environment. Proceedings of the Human Performance, Situation Awareness and Automation: User Centered Design for the New Millennium Conference, 262–267, 2000.

3.2.8 Situation Awareness Control Room Inventory

General description – The Situation Awareness Control Room Inventory (SACRI) was developed to assess SA of operators in nuclear power plants. It has four sets of questions. The first set compares the current situation with that of the recent past (19 questions for the primary circuit and 20 questions for the secondary circuit). The second set of questions compares the current situation with normal operations but uses the same questions as the first set. The third set requires the operator to predict the future state of the system over the next few minutes. Again, the same questions are used but are rated as: "(1) increase/same, (2) decrease/same, (3) increase/same/decrease, and (4) increase in more than one/increase in one/same/decrease in one/decrease in more than one/drift in both directions" (Hogg et al., 1995, p. 2413).

Strengths and limitations – Hogg et al. (1995) reported the results of four nuclear power plant simulator studies and concluded that SACRI provides a sensitive measure of SA but requires experienced operators.

Data requirements – Participant responses are compared to actual plant state.

Thresholds – Not stated.

Source

Hogg, D.N., Folles, K., Strand-Volden, F., and Torralba, B. Development of a situation awareness measure to evaluate advanced alarm systems in nuclear power plant control rooms. *Ergonomics* 38(11): 2394–2413, 1995.

3.2.9 Situational Awareness Rating Technique (SART)

General description – The Situational Awareness Rating Technique (SART) (see Figure 3.4) (Taylor, 1990) is a questionnaire method that concentrates on measuring the operator's knowledge in three areas: (1) demands on attentional resources, (2) supply of attentional resources, and (3) understanding of the situation (see Figures 3.3 and 3.4 and Table 3.6). The reason that SART measures three different components (there is also a 10-dimensional version) is that the SART developers proposed that, like workload, SA is a complex construct; therefore, to measure SA in all its aspects, separate measurement dimensions are required. Because information processing and decision making are inextricably bound with SA (since SA involves primarily cognitive rather than physical workload), SART has been tested in the context of Rasmussen's Model of skill-, rule-, and knowledge-based behavior. Selcon and Taylor (1989) conducted separate studies looking at the relationship between SART and rule- and knowledge-based decisions, respectively. The results showed that SART ratings appear to provide diagnosticity in that they were significantly related to performance measures of the two types of decision making.

Strengths and limitations – SART is a subjective measure and, as such, suffers from the inherent reliability problems of all subjective measures. Taylor and Selcon (1991) state: "There remains considerable scope for scales development, through description improvement, interval justification and the use of conjoint scaling techniques to condense multi-dimensional ratings into

		LOW						HIGH
		1	2	3	4	5	6	7
Demand	Instability of Situation							
	Variability of Situation							
	Complexity of Situation							
Supply	Arousal							
	Spare Mental Capacity							
	Concentration							
	Division of Attention							
Under	Information Quantity							
	Information Quality							
	Familiarity							

FIGURE 3.4
SART scale.

TABLE 3.6
Definitions of SART Rating Scales

Demand on Attentional Resources
Instability: Likelihood of situation changing suddenly
Complexity: Degree of complication of situation
Variability: Number of variables changing in situation
Supply of Attentional Resources
Arousal: Degree of readiness for activity
Concentration: Degree of readiness for activity
Division: Amount of attention in situation
Space capacity: Amount of attention left to spare for new variables
Understanding of the Situation
Information quantity: Amount of information received and understood
Information quality: Degree of goodness of information gained

From Taylor and Selcon (1991, p. 10).

a single SA score" (p. 11). These authors further state that "The diagnostic utility of the Attentional Supply constructs has yet to be convincingly demonstrated" (p. 12).

The strengths are that SART is easily administered and was developed in three logical phases: (1) scenario generation, (2) construct elicitation, and (3) construct structure validation (Taylor, 1989). SART has been prescribed for comparative system design evaluation (Taylor and Selcon, 1991). It has been used in aviation, cybersecurity, military operations, submarines, and surface transportation. It has also been adapted to measure cognitive compatibility.

Aviation. SART is sensitive to differences in performance of aircraft attitude recovery tasks and learning comprehension tasks (Selcon and Taylor, 1991; Taylor and Selcon, 1990). SART is also sensitive to pilot experience (Selcon et al., 1991), timeliness of weather information in a simulated flight task (Bustamante et al., 2005), and field of view and size of a flight display in a fixed-base flight simulator (Stark et al., 2001). Wilson et al. (2002) reported a significant difference in SART scores associated with alternate Head Up Display symbologies. Their participants were 27 pilots performing taxi maneuvers in a fixed-based simulator. Boehm-Davis et al. (2010) reported significantly less awareness for pilots in using Data Comm than using voice. The participants were 24 transport pilots. The data were collected in a simulator. Burke et al. (2016) used SART to evaluate the SA of 12 pilots using a trajectory optimization system in a Piaggio Avanti P180.

Selcon et al. (1992) used SART to evaluate the effectiveness of visual, auditory, or combined cockpit warnings. Demand was significantly greater for the visual than for the auditory or the combined cockpit warnings. Neither supply nor understanding ratings were significantly different, however, across these conditions. Similarly, Selcon et al. (1996) reported significantly higher

SART scores when a launch success zone display was available to pilots during a combat aircraft simulation. Understanding, information quantity, and information quality were also significantly higher with this display. There were no effects on Demand or Supply ratings. Strybel et al. (2007) reported that nine instrument-rated pilots responded to seven probe questions after a simulated ILS approach. There were significant predictions from responses to three of these questions and SART combined score, SART understanding, and SART demand.

Vidulich et al. (1995) reported increased difficulty of a PC-based flight simulation increased the Demand Scale on the SART but not the Supply or Understanding Scales. Providing additional information increased the Understanding Scale but not the Demand or Supply Scales. Crabtree et al. (1993) reported that the overall SART rating discriminated SA in a simulated air-to-ground attack while a simple overall SA rating did not. The test-retest reliability of the overall SART was not significant. See and Vidulich (1997) reported significant effects of target and display type on SART. The combined SART as well as the supply and understanding scales were significantly correlated to workload ($r = -0.73$, -0.75, and -0.82, respectively).

In a large experiment (80 men and 180 women), Fink and Major (2000) compared SART and SALIENT related to a helicopter flight simulation game (Werewolf v Comanche). The authors reported that the SALIENT had the better measurement characteristics but SART enabled them to assess the interaction between the game and the operator. In another helicopter (Blackhawk Helicopter simulator) Casto and Casali (2010) reported a significant effect of workload manipulation (decreasing visibility, increasing number of maneuvers, and number of communications) and communication signal quality on SART ratings. Brown and Galster (2004) reported no significant effect of varying the reliability of automation in a simulated flight task on SART. Their participants were eight male pilots.

Verma et al. (2010) used SART to evaluate temporal separation between commercial aircraft as well as straight in and slewed approaches. Their participants were three retired commercial pilots. Due to the small amount of data, no statistical analyses were performed. However, the authors concluded that there were no differences between 5 and 10 second separation or type of path.

In an Air Traffic Control scenario, Vu et al. (2009) reported significantly higher SA in a low air traffic density scenario than in a high traffic density scenario. Taylor et al. (1995) reported significant differences in both 3-D and 10-D SART ratings in an ATC task. Only three scales of the 10-D SART did not show significant effects as the number of aircraft being controlled changed (Information Quantity, Information Quality, and Familiarity). Durso et al. (1999) reported that Air Traffic Controllers made little distinction between understanding and SA.

Gregg et al. (2012) evaluated conflict detection automation among 13 recently retired Air Traffic Controllers using an ATC simulator. Participants were divided into two groups based on their rating of the complexity of four automation systems. Group 1 had significantly higher SART ratings for every study condition than Group 2.

Cyber Security. In a cyber threat detection task, Giacobe (2013) found no significant effects of experience on SART ratings but there were significant differences between text and graphic presentation of information on three SART scales: Attention Division, Information Quantity, and Familiarity.

Military Operations. Schuster et al. (2012) reported a significant interaction in SART scores such that a high information first group had higher SART scores on the third mission than on the second mission. The task was to identify the affiliation of soldiers in a simulated Military Operations in Urban Terrain (MOUT). The task was done in human-robot teams.

Submarines. Huf and French (2004) reported SART indicating low understanding of the situation in a Virtual Submarine Environment while SAGAT indicated high understanding as evidenced by correct responses. Loft et al. (2015) compared SPAM, SAGAT, ATWIT, NASA TLX, and SART ratings from 117 undergraduates performing three submarine tasks: contact classification, closest point of approach, and emergency surface. SPAM did not significantly correlate with SART but did with ATWIT and NASA TLX.

Surface Transportation. Read and Sallam (2017) used SART to compare SA in a Virtual Reality Head-Mounted Display, real world driving, and a flat screen to train drivers. They reported no significant differences among the three training environments.

Cognitive Compatibility. SART was modified to measure cognitive compatibility (CC-SART) to assess the effects of color coding in the Gripen fighter aircraft (Derefeldt et al., 1999). CC-SART has three primary and 10 subsidiary scales. Only the three primary scales were used: (1) level of processing, (2) ease of reasoning, and (3) activation of knowledge. Participants were seven Swedish fighter pilots performing a simulated tracking of an adversarial aircraft using the Head Up Display and detecting a target on the head down display. The CC-SART Index was lowest for the monochromatic displays and highest for the dichrome color displays. However, the shortest RT to detecting the head down target occurred with polychromatic displays. Parasuraman et al. (2009) used CC-SART to evaluate three automation modes (manual, static automation, automated target recognition) on supervision of multiple uninhabited vehicles. These authors reported that the score for level of processing was significantly lower than ease of reasoning.

Data requirements – Data are on an ordinal scale; interval or ratio properties cannot be implied.

Thresholds – Not stated.

Sources

Boehm-Davis, D.A., Gee, S.K., Baker, K., and Medina-Mora, M. Effect of party line loss and delivery format on crew performance and workload. Proceedings of the Human Factors and Ergonomics Society 54th Annual Meeting, 126–130, 2010.

Brown, R.D., and Galster, S.M. Effects of reliable and unreliable automation on subjective measures of mental workload, situation awareness, trust and confidence in a dynamics flight task. Proceedings of the Human Factors and Ergonomics Society 48th Annual Meeting, 147–151, 2004.

Burke, K.A., Wing, D.J., and Haynes, M. Flight test assessments of pilot workload, system usability, and situation awareness of TASAR. Proceedings of the Human Factors and Ergonomics Society 60th Annual Meeting, 61–65, 2016.

Bustamante, E.A., Fallon, C.K., Bliss, J.P., Bailey, W.R., and Anderson, B.L. Pilots' workload, situation awareness, and trust during weather events as a function of time pressure, role assignment, pilots' rank, weather display, and weather system. *International Journal of Applied Aviation Studies* 5(2): 348–368, 2005.

Casto, L.K.L., and Casali, J.G. Effect of communications headset, hearing ability, flight workload, and communications signal quality on pilot performance in an Army Black Hawk helicopter simulator. Proceedings of the Human Factors and Ergonomics Society 54th Annual Meeting, 80–84, 2010.

Crabtree, M.S., Marcelo, R.A.Q., McCoy, A.L., and Vidulich, M.A. An examination of a subjective situational awareness measure during training on a tactical operations trainer. Proceedings of the 7th International Symposium on Aviation Psychology, 891–895, 1993.

Derefeldt, G., Skinnars, Ö., Alfredson, J., Eriksson, L., Andersson, P., Westlund, J., Berggrund, U., Holmberg, J., and Santesson, R. Improvement of tactical situation awareness with color-coded horizontal-situation displays in combat aircraft. *Displays* 20(4): 171–184, 1999.

Durso, F.T., Hackworth, C.A., Truitt, T.R., Crutchfield, J., Nikolic, D., and Manning, C.A. *Situation Awareness as a Predictor of Performance in En Route Air Traffic Controllers (DOT/FAA/AM-99/3)*. Washington, DC: Office of Aviation Medicine, January 1999.

Fink, A.A., and Major, D.A. Measuring situation awareness: A comparison of three techniques. Proceedings of the 1st Human Performance, Situation Awareness and Automation: User-Centered Design for the New Millennium, 256–261, 2000.

Giacobe, N.A. A picture is worth a thousand words. Proceedings of the Human Factors and Ergonomics Society 57th Annual Meeting, 172–176, 2013.

Gregg, S., Martin, L., Homola, J., Lee, P., Mercer, J., Brasil, C., Cabrall, C., and Lee, H. Shifts in air traffic controllers' situation awareness during high altitude mixed equipage operations. Proceedings of the Human Factors and Ergonomics Society 56th Annual Meeting, 95–99, 2012.

Huf, S., and French, H.T. Situation awareness in a networked virtual submarine. Proceedings of the Human Factors and Ergonomics 48th Annual Meeting, 663–667, 2004.

Loft, S., Bowden, V., Braithwaite, J., Morrell, D.B., Huf, S., and Durso, F.T. Situation awareness measures for simulated submarine track management. *Human Factors* 57(2): 298–310, 2015.

Parasuraman, R., Cosenzo, K.A., and Visser, E.D. Adaptive automation for human supervision of multiple uninhabited vehicles: Effects on change detection, situation awareness, and mental workload. *Military Psychology,* 21(2): 270–297, 2009.

Read, J.M., and Sallam, J.J. Task performance and situation awareness with a virtual reality head-mounted display. Proceedings of the Human Factors and Ergonomics 61st Annual Meeting, 2105–2109, 2017.

Schuster, D., Keebler, J.R., Jentsch, F., and Zuniga, J. Comparison of SA measurement techniques in a human-robot team task. Proceedings of the Human Factors Society 56th Annual Meeting, 1713–1717, 2012.

See, J.E., and Vidulich, M.A. Assessment of computer modeling of operator mental workload during target acquisition. Proceedings of the Human Factors and Ergonomics Society 41st Annual Meeting, 1303–1307, 1997.

Selcon, S.J., Hardiman, T.D., Croft, D.G., and Endsley, M.R. A test-battery approach to cognitive engineering: To meta-measure or not to meta-measure, that is the question!. Proceedings of the Human Factors and Ergonomics Society 40th Annual Meeting, 228–232, 1996.

Selcon, S.J., and Taylor, R.M. Evaluation of the situational awareness rating technique (SART) as a tool for aircrew systems design. AGARD Conference Proceedings No. 478, Neuilly-sur-Seine, France, 1989.

Selcon, S.J., and Taylor, R.M. Decision support and situational awareness. In R.M. Taylor (Ed.) *Situational Awareness in Dynamic Systems (IAM Report 708).* Farnborough, UK: Royal Air Force Institute of Aviation Medicine, 1991.

Selcon, S.J., Taylor, R.M., and Koritsas, E. Workload or situational awareness?: TLX vs. SART for aerospace systems design evaluation. Proceedings of the Human Factors Society 35th Annual Meeting, 62–66, 1991.

Selcon, S.J., Taylor, R.M., and Shadrake, R.A. Multi-modal: Pictures, words, or both? Proceedings of the Human Factors Society 36th Annual Meeting, 57–61, 1992.

Stark, J.M., Comstock, J.R., Prinzel, L.J., Burdette, D.W., and Scerbo, M.W. A preliminary examination of situation awareness and pilot performance in a synthetic vision environment. Proceedings of the Human Factors and Ergonomics Society 45th Annual Meeting, 40–43, 2001.

Strybel, T.Z., Vu, K.L., Dwyer, J.P., Kraft, J., Ngo, T.K., Chambers, V., and Garcia, F.P. Predicting perceived situation awareness of low altitude aircraft in terminal airspace using probe questions. *Human-Computer Interaction. Interaction Design and Usability. Lecture Notes in Computer Science* (volume 4550). Berlin: Springer, July 2007.

Taylor, R.M. Situational awareness rating technique (SART): The development of a tool for aircrew systems design. Proceedings of the NATO Advisory Group for Aerospace Research and Development (AGARD) Situational Awareness in Aerospace Operations Symposium (AGARD-CP-478), October 1989.

Taylor, R.M. *Situational Awareness: Aircrew Constructs for Subject Estimation (IAM-R-670).* Farnborough, UK: Institute of Aviation Medicine, 1990.

Taylor, R.M., and Selcon, S.J. Understanding situational awareness. Proceedings of the Ergonomics Society's 1990 Annual Conference, 105–111, 1990.

Taylor, R.M., and Selcon, S.J. Subjective measurement of situational awareness. In R.M. Taylor (Ed.) *Situational Awareness in Dynamic Systems (IAM Report 708)*. Farnborough, UK: Royal Air Force Institute of Aviation Medicine, 1991.
Taylor, R.M., Selcon, S.J., and Swinden, A.D. Measurement of situational awareness and performance: A unitary SART index predicts performance on a simulated ATC task. Proceedings of the 21st Conference of the European Association for Aviation Psychology. Chapter 41, 1995.
Verma, S., Lozito, S.C., Ballinger, D.S., Trot, G., Hardy, G.H., Panda, R.C., Lehmer, R.D., and Kozon, T.E. *Preliminary Human-in-the-Loop Assessment of Procedures for Very-Closely-Spaced Parallel Runways (NASA/TM-2010-216026)*. Moffett Field, California: NASA Ames Research Center, April 2010.
Vidulich, M.A., McCoy, A.L., and Crabtree, M.S. The effect of a situation display on memory probe and subjective situational awareness metrics. Proceedings of the 8th International Symposium on Aviation Psychology, 765–768, 1995.
Vu, K.L., Minakata, K., Nguyen, J., Kraut, J., Raza, H., Battiste, V., and Strybel, T.Z. Situation awareness and performance of student versus experienced air traffic controllers. In M.J. Smith and G. Salvendy (Eds.) *Human Interface, Part II, HCII 2009*, LNCS5618 (pp. 865–874), 2009.
Wilson, J.R., Hooey, B.L., Foyle, D.C., and Williams, J.L. Comparing pilots' taxi performance, situation awareness and workload using command-guidance, situation-guidance and hybrid head-up display symbologies. Proceedings of the Human Factors and Ergonomics Society 46th Annual Meeting, 16–20, 2002.

3.2.10 Situational Awareness Subjective Workload Dominance

General descriptions – The Situation Awareness Subjective Workload Dominance Technique (SA-SWORD) uses judgment matrices to assess SA.

Strengths and limitations – Fracker and Davis (1991) evaluated alternate measures of SA on three tasks: (1) flash detection, (2) color identification, and (3) location. Ratings were made of awareness of object location, color, flash, and mental workload. All ratings were collected using a paired-comparisons technique. Color inconsistency decreased SA and increased workload. Flash probability had no significant effects on the ratings. Ruff et al. (2000) reported significantly higher SA for management by consent automation mode than for manual control or management by exception. The task was the control of one, two, or four remotely operated vehicles. However, Snow and Reising (2000) found no significant correlation between SAGAT and SA-SWORD and further only SA-SWORD showed statistically significant effects of visibility and synthetic terrain type in a simulated flight.

Data requirements – There are three required steps: (1) a rating scale listing all possible pairwise comparisons of the tasks performed must be completed, (2) a judgment matrix comparing each task to every other task must be filled in with each participant's evaluation of the tasks, and (3) ratings must be calculated using a geometric means approach.

Thresholds – Not stated.

Sources

Fracker, M.L., and Davis, S.A. *Explicit, Implicit, and Subjective Rating Measures of Situation Awareness in a Monitoring Task (AL-TR-1991-0091).* OH: Wright-Patterson Air Force Base, October 1991.

Ruff, H.A., Draper, M.H., and Narayanan, S. The effect of automation level and decision aid fidelity on the control of multiple remotely operated vehicles. Proceedings of the 1st Human Performance, Situation Awareness and Automation: User-Centered Design for the New Millennium, 2000.

Snow, M.P., and Reising, J.M. Comparison of two situation awareness metrics: SAGAT and SA-SWORD. Proceedings of the IEA 2000/HFES 2000 Congress, vol. 3, 49–52, 2000.

3.2.11 Situational Awareness Supervisory Rating Form

General descriptions – Carretta et al. (1996) developed the Situational Awareness Supervisory Rating Form to measure the SA capabilities of F-15 pilots. The form has 31 items that range from general traits to tactical employment (Table 3.7).

Strengths and limitations – Carretta et al. (1996) reported that 92.5% of the variance in peer and supervisory ratings were due to one principal component. The best predictor of the form rating was flying experience ($r = +0.704$).

TABLE 3.7

Situational Awareness Supervisory Rating Form

Rater ID#: ____________________ **Pilot ID#:** ____________________						
Item Ratings	**Relative Ability Compared with Other F-15C Pilots**					
	Acceptable		**Good**		**Outstanding**	
	1	**2**	**3**	**4**	**5**	**6**
General Traits 1. Discipline 2. Decisiveness 3. Tactical knowledge 4. Time-sharing ability 5. Reasoning ability 6. Spatial ability 7. Flight management						
Tactical Game Plan 8. Developing plan 9. Executing plan 10. Adjusting plan on the fly						

(*Continued*)

TABLE 3.7 (CONTINUED)

Situational Awareness Supervisory Rating Form

System Operation 11. Radar 12. TEWS 13. Overall weapons system proficiency						
Communication 14. Quality (brevity, accuracy, timeliness, completeness) 15. Ability to effectively use comm/information						
Information Interpretation 16. Interpreting VSD 17. Interpreting RWR 18. Ability to effectively use AWACS/GCI 19. Integrating overall information (cockpit displays, wingman comm, controller comm) 20. Radar sorting 21. Analyzing engagement geometry 22. Threat prioritization						
Tactical Employment-BVR Weapons 23. Targeting decision 24. Fire-point selection						
Tactical Employment-Visual Maneuvering 25. Maintain track of bogeys/friendlies 26. Threat evaluation 27. Weapons employment						
Tactical Employment-General 28. Assessing offensiveness/defensiveness 29. Lookout (VSD interpretation, RWR monitoring, visual lookout) 30. Defensive reaction (chaff, flares, maneuvering, etc.) 31. Mutual support						
Overall situational awareness[a]						
Overall fighter ability						

[a] Items 1 through 31 are used for supervisory ratings. The overall fighter ability and situational awareness items are completed by both supervisors and peers. (Carretta et al., 1996, pp. 40–41).

Waag and Houck (1996) evaluated the Situational Awareness Supervisory Rating Form as one of a set of three SA rating scales. The other two were peer and self-ratings. The data were collected from 239 F-15C pilots. Reliabilities on the three scales ranged from +0.97 to +0.99. Inter-rater reliability was +0.84. Correlations between supervisory and peer ratings ranged from +0.85 to +0.87. Correlations with the self-report were smaller (+0.50 to +0.58). In an earlier publication these authors referred to the scales as the Situational Awareness Rating Scales (SARS) (Waag and Houck, 1994).

Data requirements – Supervisors and peers must make the rating.

Sources

Carretta, T.R., Perry, D.C., and Ree, M.J. Prediction of situational awareness in F-15 pilots. *The International Journal of Aviation Psychology* 6(1): 21–41, 1996.

Waag, W.L., and Houck, M.R. Tools for assessing situational awareness in an operational fighter environment. *Aviation, Space, and Environmental Medicine* 65(5, Suppl.): A13–A19, 1994.

Waag, W.L., and Houck, M.R. Development of criterion measures of situation awareness for use in operational fighter squadrons. Proceedings of the Advisory Group for Aerospace Research and Development Conference on Situation Awareness Limitation and Enhancement in the Aviation Environment (AGARD-CP-575). AGARD, Neuilly-sur-Seine, France, 8-1–8-8, January 1996.

3.3 Simulation

Shively et al. (1997) developed a computational model of SA. The model has three components: (1) situational elements, i.e., parts of the environment that define the situation, (2) context-sensitive nodes, i.e., semantically-related collections of situational elements, and (3) a regulatory mechanism that assesses the situational elements for all nodes.

See and Vidulich (1997) reported that a Micro Saint model of operator SA during a simulated air-to-ground mission matched SART predictions with the closest correlation with the understanding scale of the SART.

Sources

See, J.E., and Vidulich, M.A. Assessment of computer modeling of operator mental workload during target acquisition. Proceedings of the Human Factors and Ergonomics Society 41st Annual Meeting, 1303–1307, 1997.

Shively, R.J., Brickner, M., and Silbiger, J. A computational model of situational awareness instantiated in MIDAS (Man-machine Integration Design and Analysis). Proceedings of the 9th International Symposium on Aviation Psychology, 1454–1459, 1997.

List of Acronyms

3D	Three Dimensional
a	number of alternatives per page
AET	Arbeitswissenschaftliches Erhebungsverfahren zur Tatigkeitsanalyze
AGARD	Advisory Group for Research and Development
AGL	Above Ground Level
AHP	Analytical Hierarchy Process
arcmin	arc minute
ATC	Air Traffic Control
ATWIT	Air Traffic Workload Input Technique
AWACS	Airborne Warning And Control System
BAL	Blood Alcohol Level
BVR	Beyond Visual Range
c	computer response time
C	Centigrade
CARS	Crew Awareness Rating Scale
CC-SART	Cognitive Compatibility Situational Awareness Rating Scale
cd	candela
CLSA	China Lake Situational Awareness
cm	centimeter
comm	communication
C-SWAT	Continuous Subjective Workload Assessment Technique
CTT	Critical Tracking Task
d	day
dBA	decibels (A scale)
dBC	decibels (C scale)
EAAP	European Association of Aviation Psychology
F	Fahrenheit
FOM	Figure of Merit
FOV	Field of View
ft	Feet
GCI	Ground Control Intercept
Gy	Gravity y axis
G_z	Gravity z axis
h	hour
HPT	Human Performance Theory
HSI	Horizontal Situation Indicator
HUD	Head Up Display
Hz	Hertz
i	task index

ILS	Instrument Landing System
IMC	Instrument Meteorological Conditions
in	inch
ISA	Instantaneous Self Assessment
ISI	Inter-stimulus interval
j	worker index
k	key-press time
kg	kilogram
kmph	kilometers per hour
kn	knot
KSA	Knowledge, Skills, and Ability
LCD	Liquid Crystal Display
LED	Light Emitting Diode
LPS	Landing Performance Score
m	meter
m^2	meter squared
mg	milligram
mi	mile
min	minute
mm	millimeter
mph	miles per hour
msec	milliseconds
MTPB	Multiple Task Performance Battery
nm	nautical mile
NPRU	Neuropsychiatric Research Unit
OW	Overall Workload
PETER	Performance Evaluation Tests for Environmental Research
POMS	Profile of Mood States
POSWAT	Pilot Objective/Subjective Workload Assessment Technique
PPI	Pilot Performance Index
ppm	parts per million
PSE	Pilot Subjective Evaluation
r	total number of index pages accessed in retrieving a given item
rmse	root mean square error
RT	reaction time
RWR	Radar Warning Receiver
s	second
SA	Situational Awareness
SAGAT	Situational Awareness Global Assessment Technique
SALIENT	Situational Awareness Linked Instances Adapted to Novel Tasks
SART	Situational Awareness Rating Technique
SA-SWORD	Situational Awareness Subjective Workload Dominance

SD	standard deviation
SPARTANS	Simple Portable Aviation Relevant Test Battery System
st	search time
STOL	Short Take-Off and Landing
STRES	Standardized Tests for Research with Environmental Stressors
SWAT	Subjective Workload Assessment Technique
SWORD	Subjective WORkload Dominance
t	time required to read one alternative
TEWS	Tactical Electronic Warfare System
TLC	Time to Line Crossing
TLX	Task Load Index
t_z	integration time
UAV	Uninhabited Aerial Vehicle
UTCPAB	Unified Tri-services Cognitive Performance Assessment Battery
VCE	Vector Combination of Errors
VDT	Video Display Terminal
VMC	Visual Meteorological Conditions
VSD	Vertical Situation Display
WB	bottleneck worker
WCI/TE	Workload/Compensation/Interference/Technical Effectiveness

Author Index

Subject Index